# Careers in Science and Technology:

An International Perspective

Advisory Committee
Office of Scientific and Engineering Personnel
and
Committee on International Organizations and Programs
Office of International Affairs

National Research Council

NATIONAL ACADEMY PRESS
Washington, D.C. 1995

**National Academy Press • 2101 Constitution Avenue, N.W. • Washington, D.C. 20418**

NOTICE: The project that is the subject of this report was approved by the Governing Board of the National Research Council, whose members are drawn from the councils of the National Academy of Sciences, the National Academy of Engineering, and the Institute of Medicine. The members of the committee responsible for the report were chosen for their special competencies and with regard for appropriate balance.

This report has been reviewed by a group other than the authors according to procedures approved by a Report Review Committee consisting of members of the National Academy of Sciences the National Academy of Engineering, and the Institute of Medicine.

The National Academy of Sciences is a private, nonprofit, self-perpetuating society of distinguished scholars engaged in scientific and engineering research, dedicated to the furtherance of science and technology and to their use for the general welfare. Upon the authority of the charter granted to it by the Congress in 1863, the Academy has a mandated that requires it to advise the federal government on scientific and technical matters. Dr. Bruce Alberts is president of the National Academy of Sciences.

The National Academy of Engineering was established in 1964, under the charter of the National Academy of Sciences, as a parallel organization of outstanding engineers. It is autonomous in its administration and in the selection of its members, sharing with the National Academy of Sciences the responsibility for advising the federal government. The National Academy of Engineering also sponsors engineering programs aimed at meeting national needs, encourages education and research, and recognizes the superior achievements of engineers. Dr. Harold Liebowitz is president of the National Academy of Engineering.

The Institute of Medicine was established in 1970 by the National Academy of Sciences to secure the services of eminent members of appropriate professions in the examination of policy matters pertaining to the health of the public. The Institute acts under the responsibility given to the National Academy of Sciences by its congressional charter to be an adviser to the federal government and, upon its own initiative, to identify issues of medical care, research, and education. Dr. Kenneth Shine is president of the Institute of Medicine.

The National Research Council was organized by the National Academy of Sciences in 1916 to associate the broad community of science and technology with the Academy's purposes of furthering knowledge and advising the federal government. Functioning in accordance with general policies determined by the Academy, the Council has become the principal operating agency of both the National Academy of Sciences and the National Academy of Engineering in providing services to the government, the public, and the scientific and engineering communities. The Council is administered jointly by both Academies and the Institute of Medicine. Dr. Bruce Alberts and Dr. Harold Liebowitz are chairman and vice chairman, respectively, of the National Research Council.

This material is based on work supported by the Alfred P. Sloan Foundation, the National Science Foundation, and the National Academy of Engineering.

Library of Congress Catalog Card No. 95-72621
International Standard Book Number 0-309-05427-3

Additional copies of this report are available from:

National Academy Press
2101 Constitution Avenue, N.W.
Box 285
Washington, D.C. 20055
800-624-6242
202-334-3313 (in the Washington Metropolitan Area)
B-703

Copyright 1995 by the National Academy of Sciences. All rights reserved.

Printed in the United States of America

NATIONAL RESEARCH COUNCIL

Office of Scientific and Engineering Personnel
and
Office of International Affairs

## ORGANIZING COMMITTEE FOR THE INTERNATIONAL CONFERENCE ON TRENDS IN SCIENCE AND TECHNOLOGY CAREERS

Walter A. Rosenblith, *Chairman*
Massachusetts Institute of Technology

R. Stephen Berry
University of Chicago

Ernest Jaworski
Monsanto Company
(*retired*)

Daniel A. Kleppner
Massachusetts Institute
 of Technology

John H. Moore
George Mason University

Rodney W. Nichols
The New York Academy
 of Sciences

Linda Wilson
Radcliffe College

Paolo Fasella (*ex-officio member*)
Commission of the European Communities

### Project Staff

Pamela Ebert Flattau
Director, Studies and Surveys Unit
Office of Scientific and
 Engineering Personnel
(*until April 7, 1995*)

Patrice Legro
Program Officer
Office of International Affairs
(*until October 13, 1993*)

Patricia A. Kirchner
Project Assistant
Office of Scientific and
Engineering Personnel
(*until June 10, 1994*)

# INVITED CONTRIBUTORS

PAUL BALTES, Max Planck Institute for Human Development and Education, Berlin, Germany

GLYNIS BREAKWELL, Department of Psychology, University of Surrey, Guildford, United Kingdom

DERVILLA DONNELLY, Department of Chemistry, University College Dublin, Dublin, Ireland

PIM FENGER, Ministry of Education and Science, Zoetermeer, The Netherlands

PAMELA EBERT FLATTAU, National Research Council, Washington, D.C., U.S.A.

WENDY HANSEN, Scientific, Technical, and Engineering Personnel; Industry, Science, and Technology, Ottawa, Canada

TORSTEN HUSÉN, Institute of International Education, Stockholm University, Stockholm, Sweden

KAZUO ISHIZAKA, Curriculum Research Division, National Institute for Educational Research, Tokyo, Japan

ALFRED McLAREN, Science Service, Washington, D.C., U.S.A.

JON MILLER, The Chicago Academy of Sciences, International Center for the Advancement of Scientific Literacy, Chicago, Illinois, U.S.A.

EMILIO MUÑOZ, Institute of Advanced Social Studies, Madrid, Spain

HAJIME NAGAHAMA, National Institute of Science and Technology Policy, Second Policy-Oriented Research Group, Tokyo, Japan

RICHARD PEARSON, Institute of Manpower Studies, University of Sussex, Brighton, United Kingdom

GUNNAR WESTHOLM, Organisation for Economic Co-operation and Development, Paris, France

THOMAS WHISTON, Science Policy Research Unit, University of Sussex, Brighton, United Kingdom

YU XIE, Population Studies Center, The University of Michigan, Ann Arbor, Michigan, U.S.A.

# OSEP ADVISORY COMMITTEE

LINDA WILSON, President, Radcliffe College, *Chair*

ERNEST JAWORSKI, Distinguished Science Fellow, Monsanto Company (retired), *Vice Chair*

BETSY ANCKER-JOHNSON, Vice President for Environmental Activities, General Motors (retired)

DAVID BRENEMAN, Curry School of Education, University of Virginia

DAVID L. GOODSTEIN, Vice Provost and Professor of Physics and Applied Physics, California Institute of Technology

LESTER A. HOEL, Hamilton Professor of Civil Engineering, University of Virginia

JUANITA M. KREPS, Department of Economics, Duke University

DONALD LANGENBERG, Chancellor, University of Maryland System

JUDITH S. LIEBMAN, Department of Mechanical and Industrial Engineering, University of Illinois at Champaign

BARRY MUNITZ, Chancellor, The California State University

KENNETH OLDEN, Director, National Institute of Environmental Health Sciences, National Institutes of Health

EWART A. C. THOMAS, Department of Psychology, Stanford University

ANNETTE B. WEINER, Dean, Graduate School of Arts and Sciences, New York University

WILLIAM H. MILLER, Department of Chemistry, University of California at Berkeley, *ex-officio*


*National Research Council Staff*

CHARLOTTE V. KUH, Executive Director
MARILYN J. BAKER, Associate Executive Director
KIMBERLY A. MANN, Administrative Assistant

# COMMITTEE ON INTERNATIONAL ORGANIZATIONS AND PROGRAMS

*Committee Members*

F. SHERWOOD ROWLAND, Foreign Secretary, National Academy of Sciences, *Chairman*

YVONNE C. BRILL, Aerospace Consultant, Skillman, New Jersey

GEORGE BUGLIARELLO, Chancellor, Polytechnic University

RITA R. COLWELL, President, Maryland Biotechnology Institute

ANTHONY DESOUZA, Rockville, Maryland

MOHAMED T. EL-ASHRY, Chief Executive Officer and Chairman, Global Environment Facility

HAROLD K. FORSEN, Foreign Secretary, National Academy of Engineering, *ex-officio*

EDWARD A. FRIEMAN, Director, Scripps Institution of Oceanography, *ex-officio*

LEON M. LEDERMAN, Professor of Physics, Illinois Institute of Technology, and Director Emeritus, Fermi National Accelerator Lab

THOMAS H. LEE, President, Center for Quality Management, *ex-officio*

JOHN H. MOORE, Distinguished Service Professor, George Mason University

HAROLD A. MOONEY, Paul F. Achilles Professor of Environmental Biology, Stanford University, *ex-officio*

RODNEY W. NICHOLS, Chief Executive Officer and President, New York Academy of Sciences

DAVID P. RALL, Foreign Secretary, Institute of Medicine, *ex-officio*

WALTER A. ROSENBLITH, Institute Professor Emeritus, Massachusetts Institute of Technology, *Special Advisor*

SUSAN SOLOMON, Senior Scientist, National Oceanic and Atmospheric Administration


*National Research Council Staff*

SHERBURNE B. ABBOTT, Director
LEANNA B. HANDWORK, Research Associate
EVELYN SIMEON, Office Coordinator

# ACKNOWLEDGEMENTS

Funding for the International Conference on Trends in Science and Technology Careers was provided by the U.S. National Science Foundation, the Alfred P. Sloan Foundation, and the U.S. National Academy of Engineering. In addition, the Commission of the European Communities graciously offered to host the meeting in Brussels, making it possible for additional individuals from Europe and Asia to participate in the conference.

In addition to the conference participants, a number of people contributed in important ways to the success of the meeting and the completion of this report. Walter Rosenblith, Emeritus Professor, Massachusetts Institute of Technology, with his wisdom and wit, shepherded the ideas and the planning staff in the development of the conference and the design of the resulting report. Also, Professor Paolo Fasella, Director General, Directorate General XII of the European Commission of the European Union, offered useful suggestions for the organization of the conference at key stages of its planning in his capacity of ex-officio member of the conference steering committee. Patrice Legro, who currently serves as a Senior Program Officer within the National Research Council's Center for Science, Mathematics, and Engineering Education, played an invaluable role in coordinating international communications, in assisting in the conference planning, and in reporting its outcomes. Likewise, Ian Perry, Coordinator of S/T Policies of Members States and CREST, Directorate General-Sciences, Research and Development of the Commission of the European Communities, was most helpful in providing logistical support in Brussels and contributed to the overall success of the meeting. Pamela Ebert Flattau, Director of the Studies and Surveys Unit within the National Research Council's Office of Scientific and Engineering Personnel, played a key role in planning the conference and overseeing the development of this volume. Patricia Kirchner, Administrative Assistant with the Studies and Surveys Unit, worked tirelessly to coordinate the production of the workshop papers. Donna Wiss Hannah and Anthony Quinn De Santis contributed significantly to the design and preparation of this volume. And finally, Jo Louise Husbands, Director of the Committee on International Security and Arms Control and Sherburne B. Abbott, Director of the Committee on International Organizations and Programs, had the task of weaving the pieces together into a final product, which Kimberly Mann led through production. To all of these people, we express our gratitude for their efforts.

LINDA WILSON
Chair
Office of Scientific and
Engineering Personnel

F. SHERWOOD ROWLAND
Chair
Office of International Affairs

# PREFACE

As we approach the start of a new century, it is apparent that science and technology (S&T) will continue to play a pivotal role in modern life. Skilled workers will be needed at all levels. Many of these workers will be drawn from the pool of students presently passing through school systems around the world; the remainder represent workers already employed who will need to maintain their technical skills if they are to be part of the S&T effort in the next century. Nations will be challenged, therefore, to formulate human resource policies that achieve an appropriate balance between the entry and turnover of workers in the S&T labor force while maintaining the productivity of those who remain employed.

There is growing concern, however, among educators and policymakers in many countries that fewer students are interested in pursuing careers in S&T than in the past. Because the S&T enterprise depends on the flow of new talent into the field, declining student interest could have a significant and unwanted impact in the long run on the size and quality of the workforce.

A number of countries have intensified their efforts to collect and analyze statistics that monitor the growth of the S&T labor force. "Career studies" have offered an especially important new dimension for analyzing talent flow. Such studies focus on the early phases of career development, such as student aspirations, educational patterns, and career choice.

Countries vary, however, with respect to their investment in S&T career studies. As a result, our understanding is quite limited of the extent to which countries experience common problems with respect to talent flow into S&T at these various stages of career formation.

Because of the potential for career studies to guide the development of the S&T workforce locally, regionally, and nationally, the U.S. National Research Council (NRC), in partnership with the European Union, convened a meeting to discuss what is known about the development of S&T careers from an international perspective. The idea for the conference generated from a resolution of the International Council of Scientific Unions (ICSU) noting concern about the apparent decline in interest of the world's younger generation in science. The NRC's Committee on International Organizations and Programs, which serves as the U.S. National Committee for ICSU, undertook the task of organizing with its national and international partners a response to this resolution into the conference, which is the subject of this report.

Over 40 experts familiar with the study of talent flow and the institutional dimensions of human resource development met for 2 days in Brussels in 1993 to discuss "career studies," to identify ways to improve data collection for decisionmaking, and to consider expanding the role of international organizations in relevant areas of policy formulation. Background papers were prepared by each participant and circulated in advance of the meeting.

This report presents those papers in a collected volume, together with a brief overview of the conference objectives and discussions. These materials indicate that many countries have experienced similar problems with respect to the flow of talent into S&T. Furthermore, many have adopted similar strategies for reversing unwanted trends. However, further work is needed to make data collection and analysis comparable from country to country before educators and policymakers have a clear understanding of the similarities and differences in "S&T career development" from an international perspective.

Many of the programs and activities described have made further progress and some have concluded since the conference was held. Thus, although we cannot assume that the situation in 1995 is necessarily the same as in 1993, the challenges of developing and sustaining careers in S&T described in this report will continue for some time. It is our hope that this volume will provide an important resource for those involved in the effort to turn the challenges into opportunities.

WALTER A. ROSENBLITH
Emeritus Professor
Massachusetts Institute of Technology

# CONTENTS

INTRODUCTION

PART I         MONITORING CAREER TRENDS IN SCIENCE AND TECHNOLOGY

                  Summary of Introductory Comments
                       *Richard Pearson*/9

                  Recent Developments in International Science and Technology Personnel Data Collection
                       *Gunnar Westholm*/11

                  The Trends in Scientific and Engineering Personnel in Japan and the Focus of Research in the National Institute of Science and Technology Policy
                       *Hajime Nagahama*/20

                  Developing Data Systems on Trends in Science and Technology Careers
                       *Glynis Breakwell*/36

PART II        ANALYZING TRENDS IN SCIENCE AND TECHNOLOGY CAREERS: THE LONGITUDINAL APPROACH

                  Summary of Introductory Comments
                       *Paul Baltes*/41

                  A Demographic Approach to Studying the Process of Becoming a Scientist/Engineer
                       *Yu Xie*/43

                  Human Resources in Science and Technology and the Less Developed Countries of Europe (EC-12)
                       *Emilio Munoz*/58

                  The Longitudinal Analysis of the Selection of Careers in Science and Technology
                       *Jon Miller*/72

PART III    ANALYZING TRENDS IN SCIENCE AND TECHNOLOGY CAREERS: FACTORS DETERMINING CHOICE

    Overview of Technical Papers
        *Wendy Hansen*/79

    Science and Technology Careers: Individual and Societal Factors Determining Choice
        *Thomas Whiston*/82

    Factors Behind Choice of Advanced Studies and Careers in Science and Technology: A Synthesis of Research in Science Education
        *Torsten Husen*/115

    Critique of Technical Papers
        *Alfred McLaren*/127

PART IV    UTILIZING POINTS OF INTERVENTION TO ENHANCE AND SUSTAIN INTEREST IN SCIENCE AND TECHNOLOGY CAREERS

    Introduction to Utilizing Points of Intervention to Enhance and Sustain Interest in Science and Technology Careers
        *Pim Fenger*/135

    Human Development of Science and Technology in Japan: From the Classroom to the Business World
        *Kazuo Ishizaka*/139

    Trends in Science and Technology Careers: Education Through Research
        *Dervilla Donnelly*/154

    Utilizing Points of Intervention: A Critique
        *Pamela Ebert Flattau*/164

PART V    INFLUENCING SCIENCE AND TECHNOLOGY CAREER TRENDS: THE ROLE OF INTERNATIONAL ORGANIZATIONS

    Profiles of Participating Organizations/169

APPENDIX A    THE CONFERENCE PROGRAM/177

APPENDIX B    PARTICIPANT ROSTER/180

# INTRODUCTION

In 1990, the 23rd General Assembly of the International Council of Scientific Unions (ICSU) passed the following resolution:

> We live in an era of unprecedented progress in science, but the attraction of science to the younger generation seems to be lessening in some countries. Where these statistics are available, they point to the danger of insufficient human resources in science and technology as the 21st Century opens. ICSU should examine, with other concerned scientific bodies, the magnitude of this problem.

The U.S. National Research Council's (NRC) Committee on International Organizations and Programs (CIOP), which serves as the U.S. National Committee for ICSU, undertook the task of organizing a response to this resolution. CIOP, in cooperation with the NRC's Office of Scientific and Engineering Personnel (OSEP), held discussions with the European Commission. These efforts led to the creation of a joint organizing committee to develop the agenda for an international conference on "Trends in Science and Technology Careers." The conference was held in Brussels, Belgium, on March 28-30, 1993. This volume provides the papers presented at the conference, along with written comments from the chairs and discussants for each conference session.

In preparing the conference, the organizers considered several key themes and trends. They recognized that understanding the factors shaping human resource development in science and in technical fields was essential for formulating effective science and technology (S&T) strategies at the national, regional, and global levels. The continued flow of scientists and engineers into research and related fields to encourage technological innovation was thus in the interest of every nation. Most advanced industrialized countries had monitored student career interests and had taken steps to assure a sustained supply of talented workers. Maintaining and strengthening the infrastructure for studies of the careers of individuals in S&T was critical to the future development of human resource policies. In the past, many international organizations had promoted the exchange of information about scientific opportunities among scientists; in the future, such organizations could play an important role in monitoring and enhancing human resources in S&T, as well as providing policy-relevant advice.

The goals of the conference were to:

- gauge the availability of international data or measures of career trends;

- assess the research base engaged in studying the careers of scientists and engineers;

- review some of the mechanisms designed to attract and sustain student interest in S&T; and

- identify methods to promote an interest in human resource development among the relevant international organizations.

Over 40 participants representing 15 countries, as well as 9 international multilateral organizations, took part in the 2-day invitational conference. The conference was generously hosted in Brussels by Professor Paolo Fasella, Director General for Science, Research, and Development (DGXII) of the Commission of the European Communities. The conference was divided into five sessions, each featuring a panel discussion addressing a major issue.

This report identifies and illustrates key contributions of worldwide research on S&T, and of the organizations committed to that research, to the development of policies for creating a strong and competitive workforce. The report reviews a wide range of studies that capture current approaches to the development of human resource policies for S&T in a number of industrialized nations. These include surveys of student attitudes toward S&T careers, intervention programs aimed at increasing the number and quality of individuals working in S&T, and databases designed to monitor the human resource system nationally and internationally.

In the two years since the conference was held, some of the studies discussed in the authored papers have concluded, some have made further progress, and new programs have been planned. For example, at its 1993 General Assembly, the ICSU adopted a resolution that has led to an initiative to develop a new global Program in Capacity Building in Science. Although this program is not specifically aimed at S&T careers, it addresses human resource development through achieving a higher level of public understanding of S&T to guide the application of science to the problems facing humanity. The goal of the program is to raise scientific literacy globally. The elements include: (1) establishing a Network for Capacity Building in Science, (2) strengthening primary school science education, (3) overcoming the geographical isolation of scientists, (4) promoting public understanding of science, and (5) presenting the case for science.

Although time has passed since the conference was held, the issues at the heart of this report, which were discussed by each of the panels, will continue to provide challenges and opportunities for researchers, statisticians, policymakers, and a wide array of institutions for years to come. These issues are briefly summarized below.

## Panel 1

### Monitoring Career Trends in Science and Technology

The first panel's goal was to identify what data and information on human resources collected at the global and national levels reveal about trends in S&T careers. A paper on the work of the Organization for Economic Cooperation and Development (OECD) discussed the global level, and a paper about Japan's National Institute for Science and Technology Policy (NISTEP) provided an example of national efforts.

The discussion of global data collection revealed that serious problems of data comparability and compatibility must be resolved. A key difficulty was the confusion arising from various definitions of "highly skilled workforce." The OECD, with the active support of the European Community and its statistical branch Eurostat, was addressing these problems in the course of revising its manual on research and development personnel.

A number of concerns at the national level emerged from the discussion of the statistics and data collected by Japan's NISTEP. Among the issues were the demographic evidence of an aging population and the potential shortage of young people to maintain the skilled workforce. Changing attitudes toward S&T among the younger generation and the need to diversify the scientific and engineering personnel supply by including women, older people, and non-Japanese intensified these problems, and the paper presented a number of models for addressing them.

During the commentary, the discussant noted that common problems related to comparability, specificity, periodicity, and structure existed in the collection of data at both national and global levels. The ultimate objective of modeling the factors that influence career choice, it was argued, should be to forecast the impact of these factors on the availability of S&T personnel as a basis for developing policies to affect the supply. Creating such models, so that the data could be useful for policy formation, was a serious challenge, however.

*INTRODUCTION*

## Panel 2

### Analyzing Trends in S&T Careers: The Longitudinal Approach

Theories of human development and aging have provided a framework for understanding how an individual moves into and through a career in S&T. European scientists have taken the lead in developing the theory, research, and analytical methods for a longitudinal approach to career analysis. The goals of this session were to summarize the advances in research, identify resources for longitudinal data surveys, examine how such surveys reflect cultural differences, and explore ways to make data compatible.

The panel provided new information on the use of longitudinal research in understanding trends in S&T careers. The first technical paper proposed a demographic model using modified Markov processes for studying how people become scientists or engineers. The paper presented a number of observations about age cohorts drawn from different longitudinal surveys in the United States. The analysis of the career pathway showed that the number of individuals in science and engineering decreased as the cohort aged, and that female representation in the U.S. science and engineering pool dropped significantly between ages 17 and 19.

The second technical paper took a regional view, based on a case study of the less developed countries of the European Community (Spain, Portugal, and Greece), to address the conceptual problems of understanding the links between human resources and job availability. The regional study of human capital required the consideration of the complex relationships among education, employment, and economic policy. The analysis was difficult because of the diversity and complexity of the indicators that needed to be considered, and because of the generally insufficient data available.

A suggestion for reconsidering methods emerged from the critique of these papers. The importance of longitudinal measurement in the study of career choice, and the need to build theories and models that reflect the changing population, were stressed. In monitoring the flow of young people into and out of the S&T careers, it was critical to identify the points when significant numbers of students left the field, and to study the reasons for those decisions. In the United States, for example, attention had recently focused on the impact of introductory college-level science courses on how students viewed science. A macro-level model like that proposed in the first technical paper, with appropriate definitions and adequate data, together with comparative demographic or flow models, would be helpful in understanding the dynamics of career choice in the context of economic development.

## Panel 3

### Analyzing Trends in S&T Careers: Factors Determining Choice

The goal of the third panel was to examine factors influencing individual decisions to pursue studies in S&T. These factors included opportunities to learn about what scientists and engineers do, the sociocultural environment, the quality of science teaching in schools, and the attractiveness and reward structure of a career in S&T. The evidence suggested that, more often than not, choices were based on a series of events and influences along the decision path rather than on a single causal factor or event.

The first technical paper proposed a sequential dependency model, which described the human decision process in terms of a series of interdependent and converging phases. The policy goals of the model were to maximize the desired number of entrants into S&T and to encourage the flowering of the best talent. Recognizing the inherent difficulty of human resource planning, the model emphasized institutional improvements and the provision of a flexible and creative alternative support system for students.

To examine further the factors influencing the choice of an S&T career, the second technical paper discussed current research on science education. The research suggested that the public image of science and of the role that science played in a technological society might be creating barriers to student interest. Science education needed to reach beyond the problems of what abilities were required, what specific competencies were taught, and what standards needed to be achieved to address the problem of general attitudes in society. Studies in recent years showed that ability in science played a role but was not the decisive factor in leading qualified students to an S&T career. Attitudes and motivations anchored in a particular cultural context were often more important. The degree of achievement-orientation, the existence of role models, and the perception of science as a "difficult" subject were all important factors. The role of gender was the focus of an increasing number of studies because participation in S&T varied greatly between males and females.

The implications of this research for education and science policy were not straightforward. For example, there may have been an epistemic problem arising from how knowledge in science was acquired, compared to other disciplines such as the humanities and arts.

In the discussion following the presentation of technical papers, some participants expressed the view that the relationship between education and jobs was critical and this had been severely affected by the recent economic events. To appeal to coming generations, S&T careers should offer a degree of stability and continuity. Surveys suggested that S&T degrees were perceived by some young people as too specialized to permit broader science career opportunities, and too demanding to undertake without reasonable assurance of employment. If governments and corporations wanted more skilled manpower, it was suggested, they needed to structure the occupational world to respond to these problems.

**Panel 4**

**Utilizing Points of Intervention to Enhance and Sustain Interest in S&T Careers**

The S&T career has been characterized as a series of decision points along which an individual may opt in or out of the career path. Carefully designed intervention efforts could influence the probability of an individual making the transition from one step to another. Three critical elements were noted: (1) the competence of teachers and faculty who stimulate and sustain student interest in S&T careers; (2) the equality of access to S&T careers; and (3) the professional competence of those who have entered the S&T workforce, for example, "the role model." The goal of Panel 4 was to explore mechanisms that affect critical decision points along the S&T career path, drawing upon the models of Japanese and European systems.

In Japan the National Institute for Educational Research (NIER) was leading a major educational reform effort. This was the third such reform, beginning with the Meiji Restoration and the opening of Japan to the West. The first reform effort adopted European and U.S. education models, but these rapidly evolved into a new national system tailored to specific Japanese needs. The second major reform took place after World War II, when the Japanese adopted a system strictly modeled on the U.S. approach to determining grade levels and identifying the number of years spent at each level (e.g., six-eight years at the primary level, four years at the secondary level).

The goals of NIER emphasized research on educational policies and practices, school curriculum and teaching methods for a nationwide system, and cooperation, both nationally and internationally, among educational research institutions. Particular efforts were made to create a competent teacher workforce and to strengthen the linkage between the requirements of private industry and the development of human resources.

In the European S&T context, it had long been appreciated that a nation could augment its competitive advantage through the skill and knowledge of its S&T workforce. To create highly qualified personnel, governments had expanded higher education based on the principles of equality of opportunity and, more recently, equality of access for women and minority groups. The underrepresentation of women in S&T careers was seen as a waste of both intellectual and economic resources. Questions about how to make science education more relevant to diverse communities were also examined. The subsequent discussion pointed out that educational policies needed to be reassessed in light of the current and anticipated shortages of skilled personnel in Europe.

Continued training for the current S&T workforce was another important issue. Creating networks within Europe was viewed as an important process for developing the careers of scientists and engineers. Research conferences to provide a framework for scientific debate were essential. A number of examples of networks and conferences, particularly those intended to create links among young scientists, were discussed.

The discussant commented that the two papers illustrated how changes in education and economic policies could ensure that those entering S&T careers would be assisted through the process and into employment. Successful intervention efforts were predicated on the occurrence of natural breaks—matriculation, education to first employment, and movement from one sector of employment to another. Special "at risk" groups, such as women and ethnic minorities, had been identified in a number of

*INTRODUCTION*

countries and targeted programs were being implemented to increase their skills. Studies of career outcomes were essential for assessing the quality and success of these intervention efforts. The remaining challenge was to link the theories of career patterns to policy formation and evaluation.

**Panel 5**

**Influencing S&T Career Trends:
The Role of International Organizations**

Panel 5 was designed as a roundtable discussion with presenters focused on the role of their organizations in enhancing the pool of scientific and technical personnel. Representatives of national and international organizations were asked to synthesize views of the previous panels and suggest new directions that might be adopted for follow-up action. Participants were also asked to consider the following questions:

- What lessons could be drawn from the discussions of the quality of data on scientific and technical personnel?

- How could the analysis of career choice be utilized effectively in structuring programs in scientific and technical education?

- What future actions and mechanisms should be considered by both researchers and organizations to further advance understanding of human resources in S&T?

Participants acknowledged that the drive for more and better data and research, as presented in the first question, varied from country to country and from region to region. Improved communication among data specialists, social scientists, and educational researchers could be highly effective in informing policy on human resource issues. This was particularly the case for the advanced data work of the OECD.

The second question generated a variety of responses. Some organizations with established programs for examining human resources in S&T, such as the European Community and the European Science Foundation, provided evidence that programs had been attempting to integrate some of the issues into their long-range planning. Others, with more traditional programs, such as the Third World Academy of Sciences and the North Atlantic Treaty Organization, were challenged by the broader issues of the science career path. In examining the relationship between lifelong learning and continued productivity, the partnership between education and industry emerged as a potential key factor.

The third question, which attempted to identify future actions and mechanisms, provided an opportunity to consider both the limits and the prospects of advancing the understanding of human resources in S&T. Participants noted that the scientific and technical communities were common stakeholders in the development of these resources. At present, however, the role of science as a factor in economic development, although clearly important, was not fully understood. More complete consideration of human resources in S&T required improved qualitative monitoring and continued nurturing of research on the science career process, supported by partnerships with organizations and industry.

As a whole, the conference highlighted steps that are needed to build, monitor, and sustain the pool of talented individuals who will advance S&T, whether locally or globally. Although the conference focused mostly on European, North American, and Japanese concerns and experiences, it is hoped that the analysis and ideas presented, and the dialogue forged between interested international organizations, will stimulate regional assessments concerned with problems in capacity building in the developing countries, as well as encourage further data and information sharing and consensus building in the industrialized nations to inform human resource policies.

# PART I

# Monitoring Career Trends in Science and Technology

# Summary of Introductory Comments

## Richard Pearson

Important efforts are under way in a number of countries to organize information about scientific and technical careers using common terminology and common data gathering techniques. The paper by Gunnar Westholm describes the specific efforts in this area by the Organization for Economic Cooperation and Development (OECD), while that by Hajime Nagahama offers a closer look at the organization of studies that have been undertaken in Japan to monitor the development of that workforce.

Together, both papers illustrate the challenges that lie ahead in any international effort to gauge the stocks and flows of scientists and engineers.

There are many important questions that must be answered before we can hope to achieve a common understanding of international trends in science and technology (S&T) careers. Some key questions include:

- Why are we interested in careers? Particular issues include concerns over shortages that delay development and innovation, surpluses of applicants leading to unemployment, and information to "plan" educational provision.

- What do we mean by "career"? Do we mean the period of 30-40 years of employment, or is the priority the critical stage of transition (perhaps 5-10 years) from education into the career path? Is a career something that has continuity and coherence or is it simply a summation of education and a series of different jobs?

- How far should we be interested in the 18-year-old and, indeed, younger students? These can be critical times when career counseling can have significant payoff in the course of promoting S&T careers.

- What do we need to know about careers? Can we "predict" careers for the future? How will this information help? Why not leave the development of the S&T workforce to market forces?

- What do we know about the different routes that lead people in different countries to undertake S&T careers?

- How do equity issues figure into the development of careers?

- How do we handle the issue of unemployment and underemployment in S&T? What advice should be given regarding careers in apparently overpopulated disciplines?

- How do such factors as international mobility and the "brain drain" phenomenon influence the career process? When we monitor S&T careers, are we interested in local market and/or global market factors?

- Do we mean the same things when we refer to S&T from country to country?

- What can we do about the paucity of data on S&T employment and careers? We need the data if we want to pursue the "planning" route. Additionally, if we want the market to work, we need the data to inform decisionmakers.

These are some of the issues that we have to consider if we are to determine our success in bringing the concept of career to bear on information gathering and analysis.

# Recent Developments in International Science and Technology Personnel Data Collection

## Gunnar Westholm

...combination of ... and human ...ineers and also ...ic development ...welfare and that ...l in the future. ...therefore of the ...systems and the ..., medium-, and ...illed personnel, ...nd. ...tional stock and ...y exists, which ...S&T personnel ...e their national ...ding industrial ...s. While these ...te a long time period, they have rarely been collected for the purposes of S&T analysis.

## UNESCO'S DATA ON THE WORLD STOCK OF S&T PERSONNEL

Perhaps the best known international statistics are those published by the United Nations Educational, Scientific, and Cultural Organization (UNESCO). These statistics, however, still present serious problems of international comparability, and the data are very aggregated and not always up-to-date.

According to UNESCO's statistics published in late 1992, the total stock of scientists, engineers, and technicians in the world in 1985 was estimated to be about 110.8 million compared to 79.2 million in 1980—an increase of broadly 40 percent. Extrapolating these figures, one may perhaps estimate that toward 1990 the world stock of S&T personnel could be at least 150 million. This is an interesting macro indicator in its own right but not very useful for S&T analysis in the absence of a more detailed breakdown of the categories involved, or placed in relation to a number of other economic and demographic variables.

The distribution by principal geographical zones of this UNESCO stock of S&T personnel is far from balanced. Around three-quarters of these scientists, engineers, and technicians are found in the developed countries, which account for about one-quarter of the world's total population. The remaining quarter of the S&T stock is found in developing/industrializing countries, which have three-quarters of the world population.

The concentration is particularly strong in North America. The United States and Canada together, with 6 percent of the world population, have some 30 percent of the world's S&T labor force. Europe (including the former Soviet Union) is also well represented, with around one-third of the S&T stock but only 16 percent of the world population.

For the Asian region, the UNESCO statistics aggregate highly industrialized countries, such as Japan and the "Dragon countries" (Hong Kong, South Korea, Singapore, and Taiwan), with industrializing countries, such as the People's Republic of China, India, and

Indonesia. The absolute S&T personnel potential in Asia is of the same magnitude as that of North America (30 percent) compared to its share in the world population of 60 percent.

Latin America and Africa are the regions least well off, with only 4 and 1.5 percent, respectively, of the world potential of S&T personnel; Latin America has 8 percent and Africa some 10-11 percent of the world population.

## OECD EXPERIENCE OF MEASUREING R&D PERSONNEL

The Organization for Economic Cooperation and Development (OECD) has a long experience—going back to the early 1960s—of collecting data on human and financial resources devoted to research and experimental development (R&D) activities, but, though essential, R&D is only one of a variety of S&T activities.

The OECD guidelines for the measurement of R&D, the so-called Frascati manual, are also used by UNESCO and the Statistical Office of the European Commission (Eurostat). These two international agencies are also directly involved in the ongoing work of developing a human resources manual on S&T (see below).

All OECD statistics are collected within an internationally accepted statistical framework and, if necessary, adjusted before publication to assure international comparability. For such comparisons it is necessary to work at the level of the lowest common denominator, which, in practice, may be at a rather aggregated level. For specific analytical purposes, where very detailed micro data are needed, most OECD (and UNESCO) data may well be too macro.

## EFFORTS TO COLLECT S&T PERSONNEL DATA AT THE OECD

For a variety of reasons—one being shared responsibility between directorates—the OECD has never seriously embarked on any regular collection of direct S&T personnel data, only of proxy data such as education and selected labor statistics.

S&T personnel data have been gathered on an ad hoc basis over the years, on the occasion of specific policy studies, such as those of the "brain-drain" in the late 1960s and early 1970s, the involvement of women in S&T, forecasts of the demand for and the supply of R&D personnel, or questions related to the aging or the mobility of the R&D workforce.

Interest in S&T personnel measurement has, however, remained unchanged, with jealous reference to work in the United States and the impressive and well-known data series issued in the National Science Foundation's *Science and Engineering Indicators* volumes.

A first attempt to remedy the lack of S&T personnel data was undertaken at the OECD in the early 1980s. A workshop was arranged in 1981, pushed and pulled by American initiatives, with a view to defining the conceptual framework for the international collection of such statistics. Some 30 experts from 13 countries participated. Following the discussions, the Secretariat launched a couple of experimental questionnaires requesting information on national stocks of S&T personnel, "piggy-backed" to its regular R&D surveys. The results were rather discouraging: only some four or five countries responded, and the scarce data could not be used for any serious analysis.

Perhaps the time was not ripe for this kind of data reporting. The conceptual guidelines were not satisfactory and the timing was inadequate. At that time, the principal sources of data were the national censuses, which took place at rather long time intervals. The main reason for the exceptionally low response rate was probably that the national R&D statistics experts were not themselves involved in the underlying educational and/or employment statistics, and their colleagues, in turn, were not fully motivated to participate in the exercise.

## THE OECD TECHNOLOGY-ECONOMY PROGRAM

Today, work on S&T personnel statistics is being driven by the need for information for policymakers.

The interest in the broader concept of S&T personnel was intensified on the occasion of a "Technology-Economy Program" (TEP) at the OECD (1988-1991). This was a project on policy crossing the traditional directorate borders that highlighted, among other things, the key role of human capital—once again stressing that of scientists and engineers—in the innovative process. Drawing essentially on empirical evidence, backed by scattered national statistics, the report paid specific attention to the risks of severe quantitative and qualitative skill shortages toward the end of the century, while admitting the difficulties in forecasting both the demand and the supply. (See TEP—*The Technology/Economy Programme —Technology and the Economy—The Key Relationships* OECD, Paris, 1992.)

## RECENT AND EXPECTED DEVELOPMENTS IN THE S&T WORKFORCE

Significant increases in the demand for S&T personnel in industry, services, and higher education are anticipated in many countries. Specific worries are expressed concerning difficulties in the higher education sector, which saw a rapid increase in the number of students and faculty in the 1960s and early 1970s as a result of the postwar baby boom and as a consequence of national social equity ambitions. These trends were favored by general economic expansion.

Massive faculty retirements may occur causing intensified recruitment needs in the sector already by the mid-1990s. If there is (as may be expected) strong parallel demand from trade and industry, with capacities for offering better salaries for the same personnel, severe recruitment problems could occur. On the other hand, new career openings will, after years of advancement blockage, be offered to younger faculty.

The TEP report discusses the demographic fall in national 18- to 24-year-old cohorts, which traditionally constitute the essential source of "new blood" into universities. From around 1980 to the mid-1990s, the number in this age group will fall between one-quarter and one-third in most of the OECD countries.

There is apparently no direct linear relationship between the number of age groups concerned and the number of new entrants into the higher education system. Studies undertaken by the Institute of Manpower Studies (IMS) in the United Kingdom have shown that the higher social classes, which broadly represent one-third of the population, traditionally account for some two-thirds of all higher education entrants. Since the birth rate propensities of these social classes appear to have decreased at a lower pace than that of the overall population, the expected reductions in the number of young people entering higher education have been considerably moderated.

## THE INCREASING ROLE OF WOMEN IN S&T

The expected fall in student numbers has also been compensated by continued increases in overall enrollment rates, largely explained by the growing participation of women (in some countries also ethnic minorities, students from abroad, and mature students).

Female students currently, or will very soon, account for at least one-half of all new entrants into the higher education system with, of course, large variations between disciplines (and countries). They are typically overrepresented in the social sciences and humanities. And while they now represent at least 50 percent of new medical students, they are underrepresented in other S&T disciplines, like engineering and physics. In most areas of S&T, female participation rates appear to fluctuate around 15-20 percent only of the total (but sometimes even more—the same is frequently true also for ethnic minority and mature student groups), and these proportions seem to remain rather stable over time. Increased participation of women in higher education will therefore probably be only marginal whereas, provided appropriate measures are taken, there are many reasons to believe that their interest in S&T could be intensified.

## DECREASING GENERAL INTEREST IN S&T STUDIES AND CAREERS

The general positive trend in numbers of students is offset by apparently declining interest in S&T subjects and careers. S&T degree recipients are increasingly attracted by nonscientific or technical professions (such as banking, business, and real estate), which offer better salaries and working conditions and perhaps higher professional and social status than traditional S&T careers.

The supply of S&T personnel is also increasing in many countries as a result of former technical training colleges being upgraded to engineering school status. Efforts are also being undertaken to convert or retrain graduates with nonscientific degrees into S&T studies and professions, though this transferability is probably much more difficult than the other way around.

The need for intensified continued or vocational training was highlighted in the TEP report. (This is one of the activities discussed in the ongoing preparation for another OECD manual on the measurement of "intangible investment.")

## FORECASTING THE SUPPLY AND DEMAND FOR S&T PERSONNEL

The TEP experts underline that future demand for scientists and engineers is still more difficult to predict than supply. In addition to direct and indirect demographic factors, such as mortality and retirement age, demand is affected by a number of uncertain variables such as economic growth, changes in the relative weight of different sectors of the economy (less manufacturing versus more private and/or public services), changes in the technology intensity (however defined) in these sectors, and changes in governmental policies (including "big" science programs), etc. At the same time, in some S&T areas there is oversupply or underemployment of qualified personnel, while in others, employers still have problems recruiting appropriate staff. Increasing international, as well as domestic, mobility of scientists and engineers between regions, professions, universities, and industries and organizations also makes forecasting supply and demand hazardous.

## THE IMPACT OF PUBLIC S&T/R&D PROGRAMS

There is also increasing interest in how the current S&T labor force is actually used. In the major OECD countries, an essential part of the state budget and much of the public R&D spending goes to defense-related programs, where many of the most highly qualified S&T personnel are engaged. Some examples include: defense R&D in 1990 represented 63 percent of public R&D funding in the United States, 43 percent in the United Kingdom, and 40 percent in France (to be compared with, for instance, 6 percent in Japan and 14 percent in Germany). These percentages, which were already lower than a few years earlier, have continued to decrease in the United States, France, and Germany (59, 35, and 11 percent respectively in 1992), remained stable in Japan, but increased slightly in the United Kingdom (45 percent in 1992). (SOURCE: OECD *Main Science and Technology Indicators—Biannual*).

Considerable reductions in defense spending are currently envisaged in all OECD countries. This will necessarily have a bearing on their needs for S&T/R&D personnel and may result in serious cuts in this labor force and, if not properly channeled, in loss of vital skills, at least until more civil-oriented employment alternatives can be found. Such conversions are hard to handle; this is one of the most crucial problems—resulting in both a domestic and international "brain-drain"—with which many of the new eastern countries are currently wrestling.

## THE OECD S&T MANUALS

The TEP report regretted the lack of accurate statistical tools for a better policy understanding of the human capital situation, and in 1991 the Scientific, Technological, and Industrial Indicators Division (STIID) of the OECD Directorate for Science, Technology, and Industry was invited to undertake the preparation of a manual measuring stocks and flows of human S&T resources. (In the autumn of 1992 the STIID was renamed the Economic Analysis and Statistics Division.)

The OECD is already a world leader in methodological developments for the collection of R&D and S&T statistics and indicators and has recently issued guidelines, in the Frascati family of manuals, on the measurement of innovation activities and the technology balance of payments. Work is under way for a patents manual and is well advanced on manuals on the measurement of "intangible investment" and high-, medium-, and low-tech activities. Corresponding databases have been established or are under development.

The OECD Directorate for Education, Employment, Labor, and Social Affairs collects educational statistics in a common OECD/UNESCO/Eurostat questionnaire, although each international agency has its own databases and database management. These third-level data on students and faculty will serve as essential inputs for the forthcoming S&T personnel databases.

The S&T manuals should provide the statistical framework for complete databases using standard international classifications wherever possible and ensure structured links to already existing databases. The guidelines should indicate how available data could, or could not, be used. The manuals are intended to meet the very diverse needs of both *producers* and *users* of data. The guidelines should be addressed for both national and international statistical purposes (including analysis). Even if, until quite recently, all these manuals were essentially directed toward the needs of economically developed countries (typically members of the OECD), there is now increasing demand from non-member countries, including the industrializing world.

Preparing such manuals is a time-consuming exercise. More time is needed for completely new statistics (such as the R&D or innovation statistics) than for the use of S&T purposes of already existing data, even if they were originally assembled for reasons other than S&T policy (i.e., the technology balance of payment and the patents statistics). The S&T personnel manual project belongs to the latter category.

The procedure is usually as follows: after theoretical discussion—sometimes starting at a very academic level—and the preparation of a draft manual, a pilot survey is undertaken. Drawing on the results of this first survey, the manual may be revised, and, if necessary, the experience renewed before regular surveys are launched and related databases built up.

## THE NEW S&T HUMAN RESOURCES MANUAL

In the case of the human S&T resources manual, the OECD had asked Dr. Richard Pearson of the IMS, who served as a consultant, to prepare the first draft. The OECD and the IMS teams began by examining available theoretical literature. They decided to draw as much as possible on already existing practical national and international experience. This was done through an ambitious inventory survey, using a rather comprehensive set of questionnaires prepared jointly by OECD and Eurostat, addressed to member countries and to UNESCO.

All this work has been fully backed by the European Commission in Brussels, in particular the DGXII, the Directorate General for Science, Research, and Development.

No less than 22 of the 24 OECD member countries responded—of course in more or less detail—to the inventory survey. This exceptionally high response rate testifies to the general interest in these S&T personnel issues, even if the results may have been biased by the composition of the group of respondents where R&D/S&T indicators experts predominated.

The preliminary conclusions of the inventory were presented by Eurostat at a first international workshop in Paris in October 1992, gathering some 70 experts from 25 OECD and non-OECD member countries, the European Commission, and UNESCO to discuss Dr. Pearson's draft manual.

The final results of the inventory were analyzed by Eurostat and served as background information at the expert meetings in Luxembourg and Paris in 1993 for revison of the manual.

Some of the principle conclusions of this exercise are presented—pell mell—below.

There is very little systematic data collection of S&T personnel, and more use is made of various kinds of proxy statistics. Only three or four countries (including the United States) have adopted some kind of a national definition of S&T personnel and collect corresponding data. R&D and educational policies appear to be the most important markets for this kind of indicator. National considerations prevail over international comparisons. Only some of the largest countries undertake systematic efforts of modeling (defined as the use of formal mathematical models for long term projections) or forecasting (understood as short or medium term outlooks based on known factors such as demographic developments and educational enrollments) of S&T personnel. Central statistical personnel data (notably education and labor statistics) are the prime source, but similar information is also drawn from associated agencies. It should be possible to extract and use sources as far as both levels of education and (broad or detailed) fields of science are concerned (including the social sciences and the humanities). Data on gender and age appear to be generally available. Information on national and international mobility (including data on nationality, etc.) appears to be rather scarce, even if several countries undertake "first destination" surveys of graduates.

## TERMINOLOGY AND DEFINITIONS

The preparatory work revealed that there is considerable confusion in the terminology employed and that these semantic descriptions also reflect broad variations in the coverage, or perception, of the data. For instance, the following denominations of S&T personnel (or manpower) were frequently used:

- highly qualified manpower

- scientific and technical or technological personnel

- scientific and engineering personnel

- scientific, technological, and engineering personnel

- scientists and technologists
- highly skilled personnel
- highly qualified technological manpower
- academic-level researchers and teachers
- research and technological development personnel

At the previously mentioned workshop it was decided to use the denomination "human resources for S&T" (HRST) for the purpose of the new manual and future work.

Only one of the above classes—scientific, technological, and engineering personnel—seems to have an internationally adopted definition, but the Frascati manual also gives definitions of various categories of R&D personnel.

In its report *Recommendation Concerning the International Standardization of Statistics on Science and Technology*, adopted by the General Conference in Paris in 1978, and in the subsequent *Manual for Statistics on Scientific and Technological Activities*, UNESCO defines scientific, technological, and engineering personnel as follows:

> Scientific and technical personnel can be defined as the total number of people participating directly in S&T activities in an institution or unit and, as a rule, paid for their service. This group should include scientists and engineers, technicians, and auxiliary personnel. . . .

Later in the manual, UNESCO also defines what is meant by the "scientific and technical manpower potential" for which two categories are identified: the "total stock of qualified manpower" and the "number of economically active qualified manpower."

The two categories are defined as follows:

> (The) *Total stock of qualified manpower* comprises the total number of persons with the necessary qualifications for personnel in categories "scientists and engineers" and "technicians," regardless of economic activity (production, S&T activities, the professions, no gainful employment, etc.), age, sex, nationality, or other characteristics present in the domestic territory of a country at a given reference date.

> (The) *Number of economically active qualified manpower* comprises the total number of persons with the necessary qualifications for personnel in categories "scientists and engineers" and "technicians" who are engaged in, or actively seeking work in, some branch of the economy at a given reference date.

The following graphic illustrates the relationship between these two UNESCO main aggregates and the scientists, engineers, and technicians engaged in S&T activities (of which R&D):

```
         Total Stock of Qualified Manpower
I-----------------------------------------------------I
                 Economically Active
         I-------------------------------------------I
                      In S&T Activities
                 I--------------------------------I
                              Of which R&D
                              I-------------I
I--------------------I--------------------------------I
I-----------I/ / / / I:/:/:/:/:/:/:/:/I/:x/:x/:x/:xI
I-----------I/ / / / I:/:/:/:/:/:/:/:/I/:x/:x/:x/:xI
```

Above, several aggregates of S&T personnel that could be measured and used for data collection and analytical purposes have been indicated. One of these aggregates may be that of "scientists and engineers," which is perhaps too frequently used as a block, with no separate distinction between the two elements.

## DEFINITIONS IN TERMS OF QUALIFICATIONS OR OCCUPATION?

Here again, several approaches are possible. Should the categories of S&T personnel be identified in terms of their formal educational background (i.e., by qualifications), or in terms of the jobs they are actually performing (i.e., by occupation)? Or perhaps by a mixed qualification/occupation criterion (for instance, when classifying people who do not possess the usual requested diplomas for the jobs they are performing but who, all the same, have the corresponding professional skills and experience)?

Given that both approaches meet specific policy needs, participants wished that both the qualifications and the occupation criteria be maintained for further conceptual work, with a slight preference for the first approach (due to better data availability).

## SHOULD TECHNICIANS BE INCLUDED?

It is, furthermore, not very clear, when we talk about scientific, technological, and engineering personnel, whether the category of technicians (or people with corresponding educational level backgrounds or professional experience) should be included. Lack of competent technical staff may create problems just as serious as an insufficient supply of scientists and engineers, and concern has been expressed by OECD, and industrializing countries as well, of risks of forthcoming "bottlenecks" in the supply of technicians.

At the workshop, participants spoke in favor, but far from unanimously, of the inclusion of technicians in the coverage of the HRST. The principal argument in favor of their exclusion was essentially one of practical measurement problems; whereas it is quite easy to define what we mean by scientists and engineers in terms of the International Standard Classification of Occupations and the International Standard Classification of Education, it is more problematic to define the technicians (and associated categories).

## WHAT DISCIPLINES SHOULD BE INCLUDED IN S&T?

Another question concerns the coverage of the science and technology activities themselves. The R&D statistics manual, which to some extent serves as a model for the HRST manual, recommends the inclusion of the natural sciences and engineering, including the agricultural and medical sciences, on the one hand, and the social sciences and humanities on the other.

In his draft manual, Dr. Pearson proposed that only the natural sciences and engineering should be covered—the social sciences and humanities should be excluded. This appeared to be a typically Anglo-Saxon position that met with resistance from countries with the "Wissenschaft" approach. Some workshop participants spoke in favor of including the social sciences but not the humanities. It was finally agreed to recommend a coverage identical with that of the R&D statistics, though with sufficiently distinct subclasses to permit the separate identification of the social sciences and humanities (to make their exclusion possible).

## ALL SKILLED PERSONNEL OR ONLY THOSE ENGAGED IN S&T?

Similar discussions took place on whether we were interested in the total potential ("reserve") of S&T personnel (to be identified as all people with appropriate educational backgrounds or working in such occupations) or only those who are economically active, and, if so, whether we should concentrate on only those who are directly involved in S&T activities (such as the second UNESCO category mentioned earlier). The latter approach would probably exclude all medical and associated practitioners and, of course, all people with S&T training but working in nonscientific activities. Here, people working in the higher education sector (i.e., the third level of the national educational system) would be included in the data, but should people with the same S&T diplomas working at lower levels of the educational system really be excluded?

## DEPRECIATION OF QUALIFICATIONS

Another directly related problem was raised. S&T are activities in rapid evolution, necessitating continuous theoretical and practical updating. Should people with a basic S&T training obtained in areas of rapid change, say 20 years ago, still be included in the S&T potential? Should retired scientists and engineers or persons older than, for example, 70 years, still be considered for the measurement of total S&T potential?

No specific recommendations were made by workshop participants on this point, but, at a later stage of the discussions, it was recommended that highest priority be given to the measurement of HRST under the age of 70.

## STOCKS AND FLOWS OF S&T PERSONNEL

The 1981 and 1992 workshops concluded that information is needed on both stocks and flows for a better understanding of the S&T potential and possible future developments. Stock data present a snapshot of the situation at a particular point in time (and how the labor force is employed), whereas flow data indicate sources of change in the stock (inflows or outflows) during a given period of time, usually a year.

## SOURCES OF DATA

As already mentioned, few countries undertake general surveys of their S&T personnel resources. Such information is drawn instead from a variety of other sources.

The most commonly cited source of stock data is the population census undertaken, depending on the country, at perhaps five- to ten-year intervals. Its principal disadvantage is that the results are often available only several years in arrears. Other sources are household surveys, surveys of industrial and R&D activities, various kinds of educational and employment surveys (usually annual), etc.

For specific policy studies, ad hoc surveys have also been undertaken, and central population and other administrative registers are increasingly used. Specialized trade and professional bodies maintain member registers that may be of interest.

For the flow data, the principal source is, of course, the educational statistics, in the short-term notably for numbers of people in the higher education pipeline, i.e., numbers of students entering, already in, or graduating from the higher education system (inputs, throughout, and outputs). First-destination surveys, which may shed light on various aspects of flows from higher education to employment, are, as indicated earlier, carried out in several OECD member countries.

The stock and flow chart below is an illustration of the principal stocks and flows of S&T personnel within the national science and technology system. (See Figure 1.)

## S&T PERSONNEL VARIABLES OF POLICY

Ideally, we should like to have gross data for all the arrows and boxes of the stock and flow chart, each cross-classified with detailed breakdowns by, for instance, type of personnel, occupation, sector of employment, educational background, broad or detailed field of qualification, gender, age, nationality, etc.

In practice, all this would be too ambitious; it would be exceedingly difficult to handle such a multidimensional matrix, and, once again, the more detail, the more difficult the international comparability! A realistic selection of possible variables must be made very shortly. This choice will be greatly influenced by the current availability of corresponding, relevant (education, employment, etc.), and internationally comparable data.

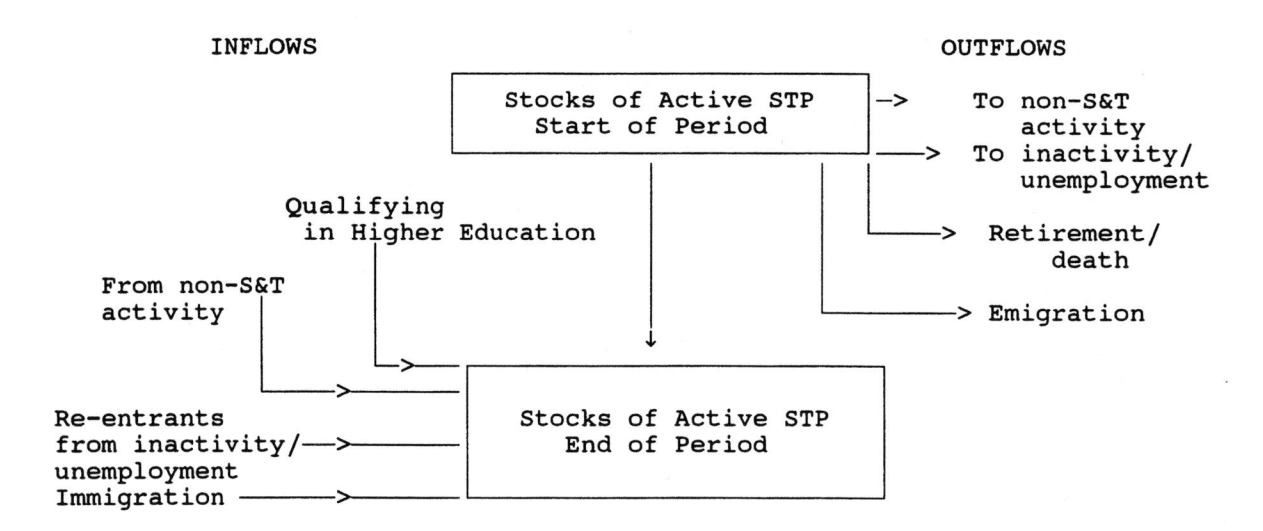

NOTE: This early stock and flow chart will be considerably amended in the final version of the HRST manual.

SOURCE: Institute of Manpower Studies, Falmer, Sussex, United Kingdom.

**FIGURE 1** Principal stocks and flows of S&T personnel.

## FULL-TIME EQUIVALENCE VERSUS HEAD-COUNT

Internationally comparable R&D personnel data have been collected in terms of full-time equivalence (FTE) only. While FTE series are considered more accurate than head-count data for international comparison of the R&D volume proper, they do not permit serious comparisons with other personnel series, such as population, education, and labor statistics. However, this is being amended. The new (1992) version of the Frascati manual recommends, for example, that R&D head-count data also be collected. This is a direct consequence of increased interest in human capital resources and constitutes the necessary linkage between the Frascati manual on R&D and the new manual under development for the measurement of total stocks and flows of S&T personnel.

## COMPATIBILITY WITH INTERNATIONAL CLASSIFICATIONS

For all these breakdowns, internationally adopted classifications should be used, such as UNESCO's International Standard Classification of Education (ISCED), the International Labor Office's International Standard Classification of Occupations, and the United Nations' International Standard Industrial Classification and System of National Accounts, of which the latter three have just been revised. Revision work has also been initiated of ISCED, which defines different levels of education and detailed fields of study classification.

Besides traditional demographic statistics needed as a basis for forecasting and modeling, a variety of other factors of importance for international comparisons of human S&T resources systems are also discussed in the draft HRST manual, such as shortages and vacancies, salaries, retirement ages, and attitudes to S&T studies and careers.

## WHAT NEXT?

Following the discussions at the 1992 workshop, work has continued with a view toward improving the concepts for measuring HRST. A first revised version of the manual, which draws to a large extent on the findings of the inventory of national and international practice and experience described earlier, was circulated to OECD, Eurostat, and UNESCO experts during the summer for another round of technical discussions in the fall of 1993. (The experts' final "green light" was given to the second revised version of the HRST manual at a meeting in Canberra, Australia, in April 1994, and it is hoped that it will be available as a "general distribution" OECD publication before the end of 1994.)

In the meantime, it is also hoped that the provisional guidelines will be tested in practical ways, notably within the framework of the ambitious project on Scientific, Technical, and Engineering Personnel at Industry, Science, and Technology Canada. OECD is taking steps to use notably some of the already available (and reasonably internationally comparable) third-level educational statistics for this purpose.

# The Trends in Scientific and Engineering Personnel in Japan and the Focus of Research in the National Institute of Science and Technology Policy

Hajime Nagahama

## ESTABLISHMENT OF NISTEP AND ITS TASKS

The National Institute of Science and Technology Policy (NISTEP) was founded in July 1988 as an organization within the Science and Technology Agency. NISTEP analyzes the increasingly complex and diverse circumstances surrounding science and technology (S&T) and carries out both theoretical and empirical studies on various subjects with an eye toward forming a S&T policy geared to the needs of a new age.[1]

Toward this end, NISTEP, in close cooperation with the Council for Science and Technology Policy and other related domestic organizations, as well as with foreign organizations involved in S&T policy, engages in research activities while positively promoting exchanges with researchers at home and abroad. (See Figure 1.)

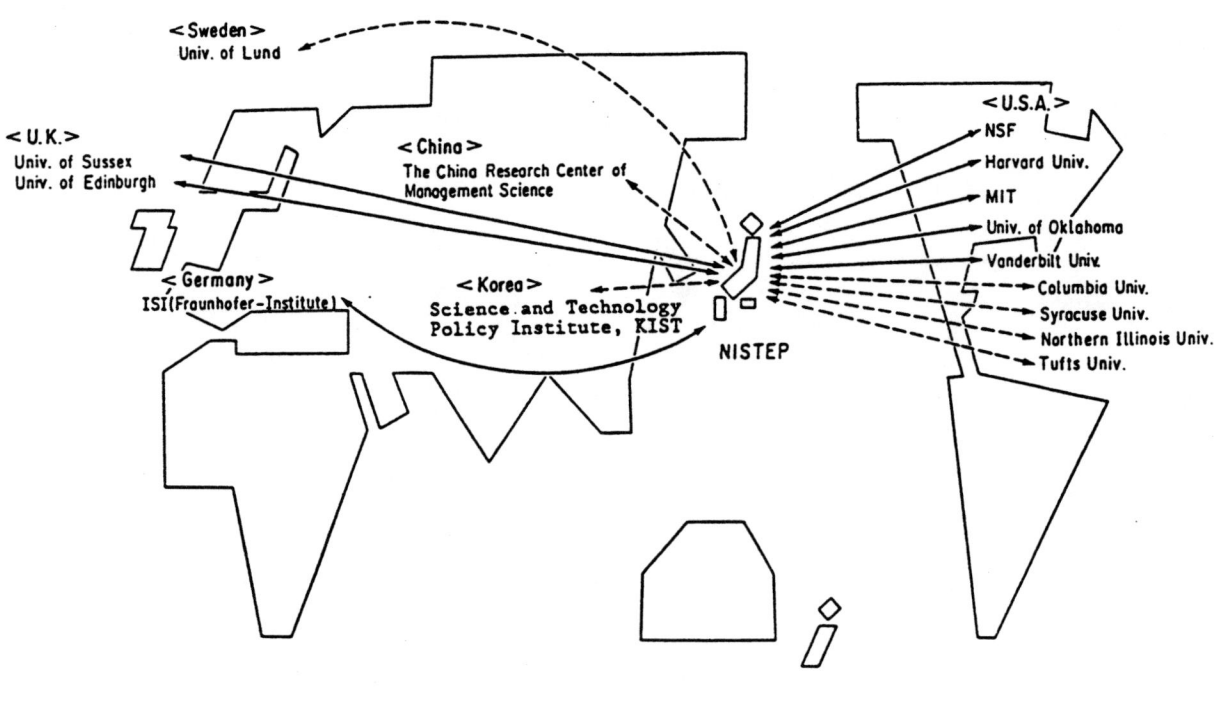

**FIGURE 1** Cooperation overseas.

## BACKGROUND OF THE PROBLEM OF SCIENTIFIC AND ENGINEERING PERSONNEL IN JAPAN

### Historical Character of Education for Scientific and Engineering Personnel

Since the Meiji Restoration, the Japanese government has imported western S&T as a combined civilization. The Meiji government sent many students to study abroad and, in parallel, employed many foreign scientists and engineers to modernize, enrich, and strengthen Japan.

Engineering education has been emphasized in the national educational policy since the Meiji era. Therefore, Japanese universities have had both science and engineering faculties and have educated both scientists and engineers since the early period. The national government has mainly taken responsibility in fostering scientific and engineering personnel in Japan.

### Current Situation in Scientific and Engineering Personnel Supply

The supply and demand gap of scientific and engineering personnel will continue both in quantity and quality. In recent years many excellent university graduates in the scientific and engineering fields have decided to work in the banking and insurance industries and not in manufacturing industries. Aware of the decline of scientific and engineering facilities in Japanese universities, people anticipate new policies aimed at promoting personnel in advanced and basic science and engineering fields.

### Problems and Prospects in the Near Future in Japan

- Shortage of young human resources and aging of demographic structure (see Figure 2)

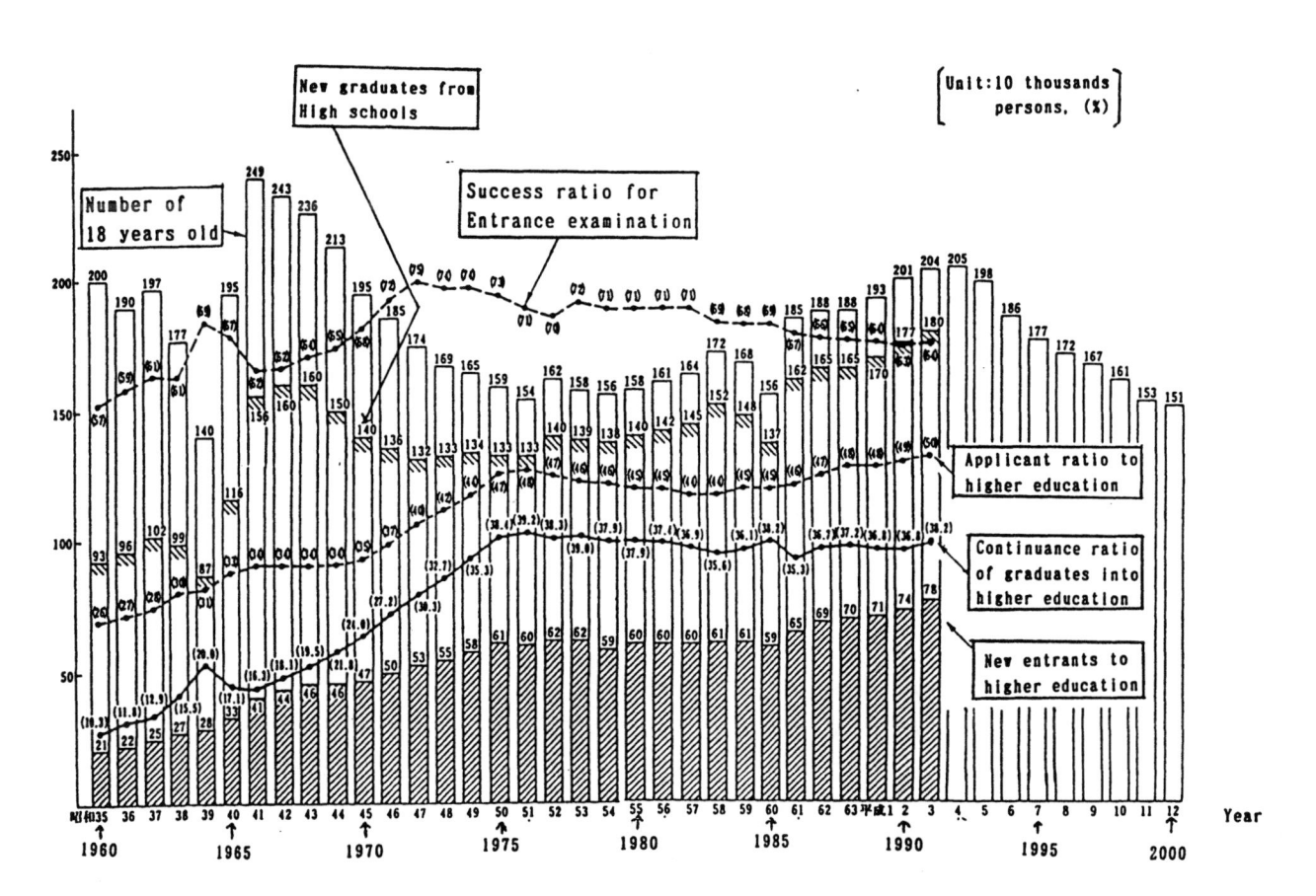

SOURCE: Planning Division, Higher Education Bureau, Ministry of Education, Science, and Culture, Japan.

**FIGURE 2** Demographic change of 18-year-olds and the scale of higher education.

- Changing attitudes of youth toward S&T (young people drift away from S&T) (see Figure 3)

- Diversification and globalization of scientific and engineering personnel supply—female, aged people, and foreigners

## MAIN CONTENTS AND FOCUSES OF RESEARCH CONCERNING SCIENTIFIC AND ENGINEERING PERSONNEL IN NISTEP

Main research concerning scientific and engineering personnel is classified into the following six groups by theme:

- Research on public understanding of S&T and STS communication

- Research on careers and professional activities of scientific and technology personnel

- Research on trends of S&T careers through S&T award

- Research on choice of field by university applicants and employment trends of science and engineering graduates

- Research on science and engineering doctorates in Japan

- The Japanese S&T indicator system

Next, I will talk about the simple summary of respective studies belonging to the above themes.

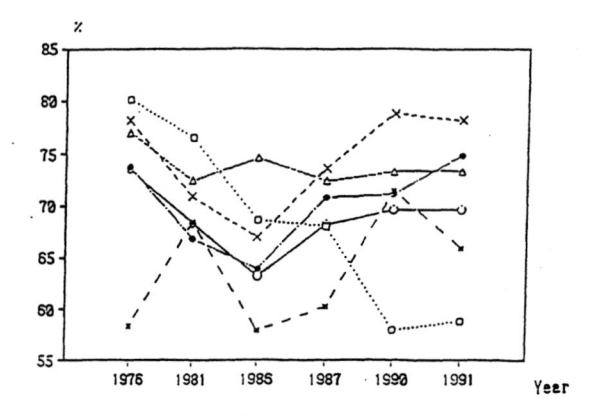

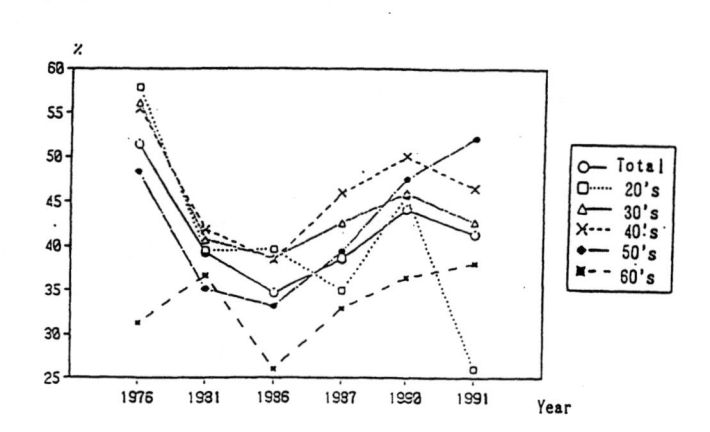

SOURCE: 1991 Survey: AC, Opinion Survey of Science and Technology
Other Surveys: Public Relations Division, Prime Minister's Office, Opinion Survey on Science, Technology, and Society, Japan.

**FIGURE 3** Changing ratio of interest in news and topics on science and technology.

## Research on Public Understanding of S&T and STS Communication

Studies belonging to this group are based on public opinion surveys toward S&T, and study the hypotheses and structure models of public attitudes and acceptance of S&T, the S&T triad, and scientific literacy. These studies aim at analyzing attitudes toward S&T, measuring scientific literacy, and encouraging smooth communication between the public, scientists and engineers, and organizations (government, private enterprises, etc.). (See Figure 4.)

The studies belonging to this group started from the study of *Science, Technology, Society, and Communication* (STSC study) (ATS Report No. 17, March 1991) by the 2nd Policy-Oriented Research Group.

This study concluded that the public, scientists and engineers, and organizations (government, private companies, etc.) have to maintain smooth communication through co-ownership of scientific and technological information in the prospective society, which is highly industrialized by advanced S&T, and proposed a S&T triad model, concepts of STCC and SCC, in order to understand their relation.[2]

The study group developed a hypothesis based on the S&T triad shown in Figure 5. This communication model represents the social relation of S&T communication.

## Research on Careers and Professional Activities of Scientific and Technology Personnel

The 2nd Policy-Oriented Research Group published a report *Public Attitudes on Science and Technology: Based on Opinion Survey Results* (NISTEP Report No. 2, June 1989) in connection with the above STSC study.

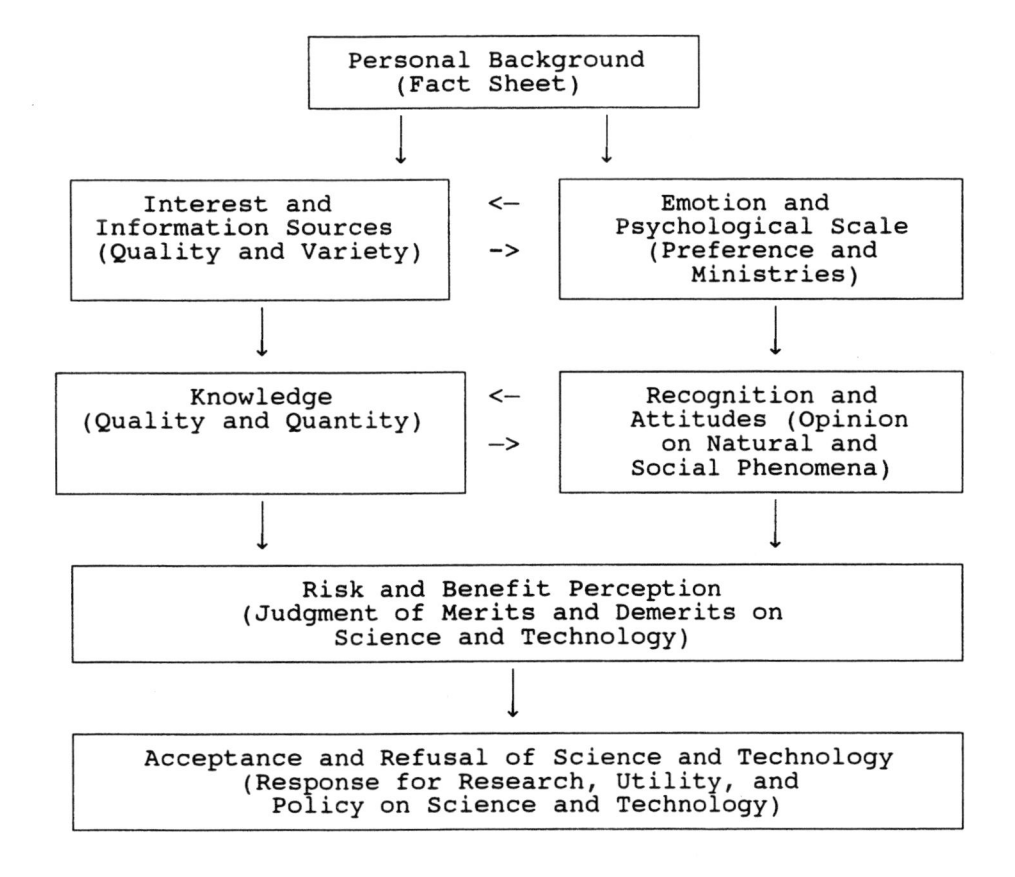

```
              Personal Background
                 (Fact Sheet)
              /              \
             ↓                ↓
   Interest and           Emotion and
   Information Sources  <—  Psychological Scale
   (Quality and Variety) —>  (Preference and
                              Ministries)
             ↓                ↓
        Knowledge          Recognition and
   (Quality and Quantity) <— Attitudes (Opinion
                          —>  on Natural and
                              Social Phenomena)
             ↓                ↓
         Risk and Benefit Perception
      (Judgment of Merits and Demerits on
              Science and Technology)
                     ↓
      Acceptance and Refusal of Science and Technology
         (Response for Research, Utility, and
              Policy on Science and Technology)
```

SOURCE: Study Group for the International Comparative Study of Public Understanding of Science and Technology, Japan.

**FIGURE 4** Structural model of public attitudes and acceptance of science and technology.

## S & T Triad

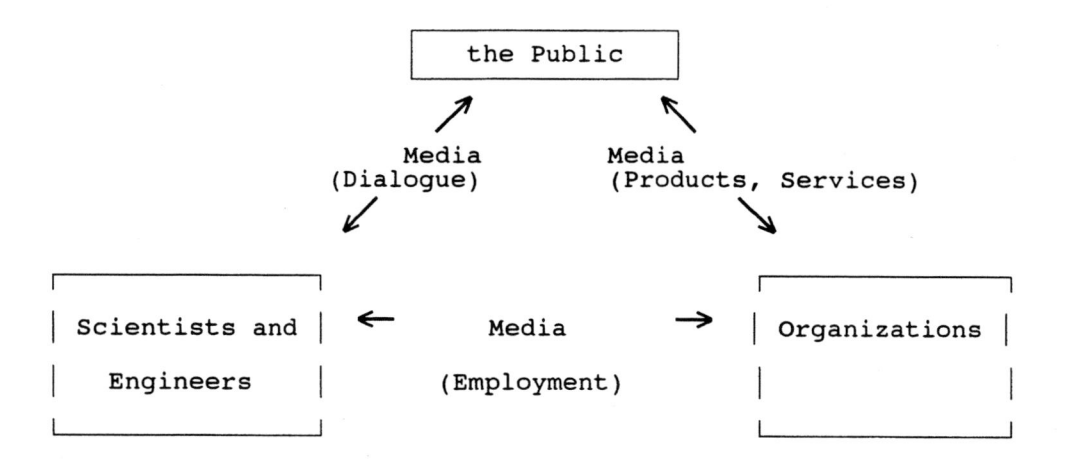

```
                    the Public
                  ↗           ↖
            Media              Media
          (Dialogue)      (Products, Services)
              ↙                   ↘
  ┌──────────────┐              ┌──────────────┐
  │ Scientists and│ ← Media →   │ Organizations│
  │   Engineers  │ (Employment) │              │
  └──────────────┘              └──────────────┘
```

SOURCE: NISTEP Report No. 17, Science, Technology, Society, and Communication.

**FIGURE 5** The communication model among the public, scientists and engineers, and organizations (S&T Triad).

Since 1990, NISTEP has participated in an international comparative study of *Public Understanding of Science and Technology*.[3]

NISTEP organized an international symposium on public understanding of S&T and science and mathematics education in Tokyo in October 1992 in cooperation with the International Council Group of the above international comparative study.

Studies belonging to this research theme aim at looking for directions to improve research conditions for Japanese scientists and engineers who are doing creative research activities.

The 2nd Policy-Oriented Research Group administered two studies under this theme. The first study resulted in a report on the *Background and Professional Activities of Natural Science Research Personnel in Japan* (NISTEP Research Material No. 20, February 1992). This report compiled the results of two surveys.

The first survey emphasized from which universities Japanese university teachers in natural science fields received their doctorates. It included university teachers in natural science fields working at all national and public universities and 10 selected private universities. The second survey showed the actual activities and professional careers of graduates. It included 678 researchers in universities, national and public laboratories, and private enterprises.

The report of the second study dealt with *The Interchange of Researchers and Engineers Between Japan and the Other Countries* (NISTEP Report No. 16 and NISTEP Research Material No. 12, March 1991). This study was based on the following three government statistics:

1. Annual Report of Statistics on Legal Migrants, Ministry of Law

2. Statistics on Foreigners Staying in Japan, Ministry of Law

3. Annual Report of Statistics on Japanese Nationals Overseas, Ministry of Foreign Affairs

The report features the actual results of the exchange of researchers and engineers between Japan and other countries for a period of 20 years (1970-1989) by object, nation, and region. (See Figures 6 and 7.)

### Research on Trends of S&T Careers Through S&T Award

This research aimed at emphasizing the research conditions and highlighting excellent research cases that were awarded by the Minister of State for S&T. The

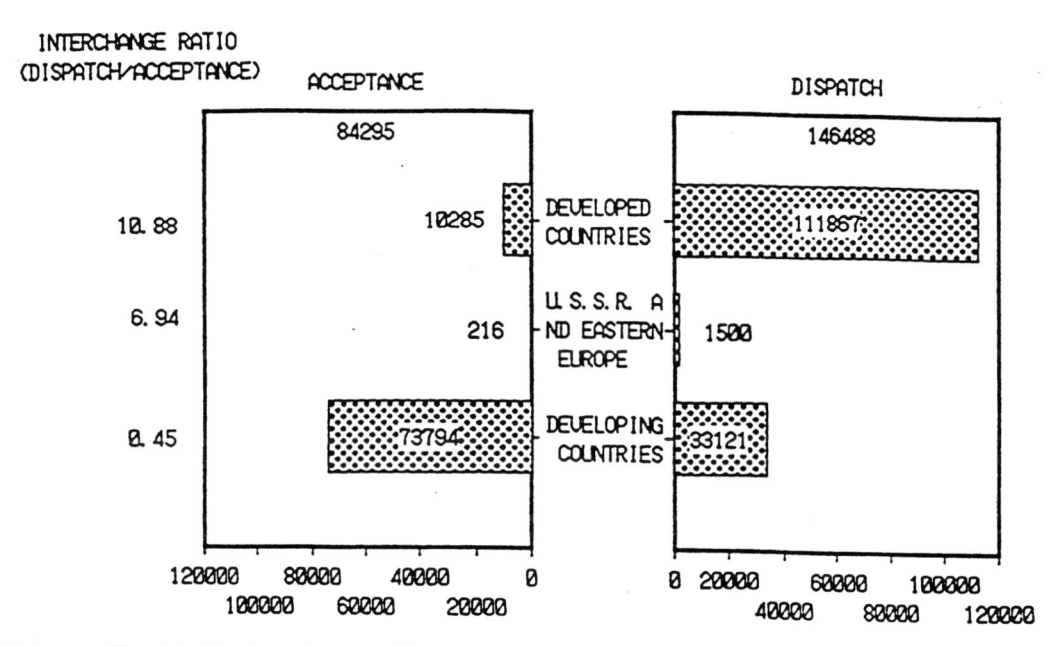

SOURCE: NISTEP Report No. 16, The Interchange of Researchers and Engineers between Japan and Other Countries: Study based on Annual Report of Statistics on Legal Migrants, Statistics on Foreigners Staying in Japan, and Annual Report of Statistics on Japanese Nationals Overseas.

**FIGURE 6** Dispatch and acceptance of researchers and engineers (1989).

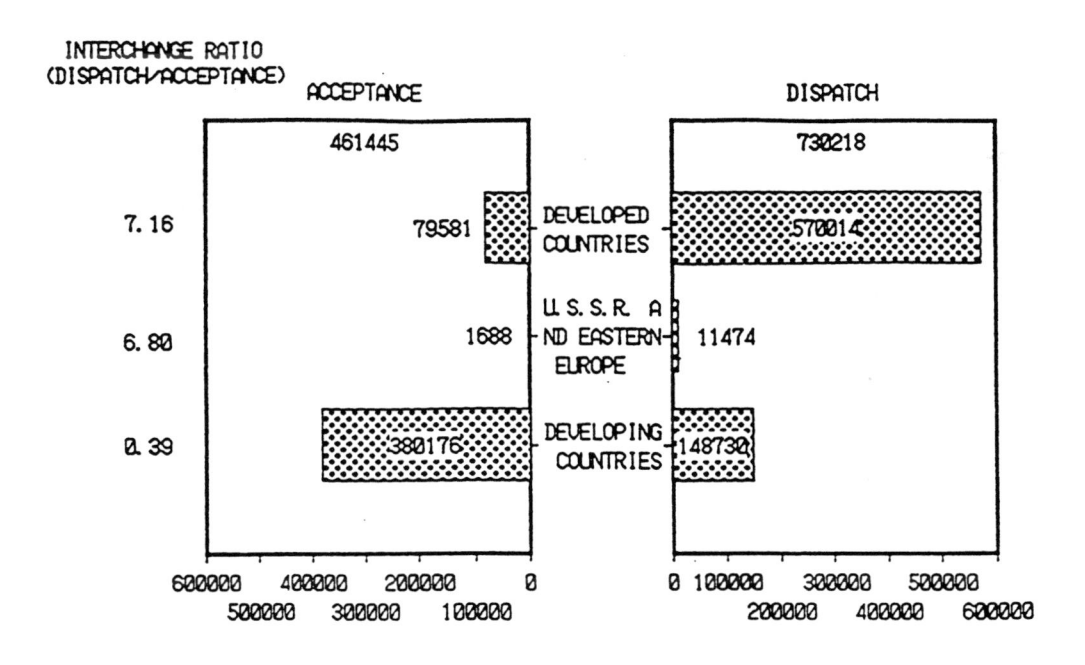

SOURCE: Report No. 16, The Interchange of Researchers and Engineers Between Japan and Other Countries: Study based on Annual Report of Statistics on Legal Migrants, Statistics on Foreigners Staying in Japan, and Annual Report of Statistics on Japanese Nationals Overseas.

**FIGURE 7** Dispatch and acceptance of researchers and engineers (20 years).

first step of this study was published as the *Trends of Science and Technology Activities in Japan Using Science and Technology Award Statistics: Persons of Scientific and Technological Merits, Commendation by the Minister of State for Science and Technology* (NISTEP Report No. 10, March 1990). In this report, 637 cases over 31 years (1959-1989) were listed and analyzed from multiple viewpoints.

This research now steps into the second round, where several remarkable cases are selected from those included in the first step of the study. Next, we are going to analyze and study the excellent achievements and the research conditions that bring about such achievement. The results of this study will be published by the end of 1993.

## Research on Choice of Field by University Applicants and Employment Trends of Science and Engineering Graduates

Three NISTEP reports and one seminar report were published in this group.

The first report considered the *Employment Trends of Science and Engineering Graduates* (NISTEP Report No. 1, June 1989). This study focused on the trend since 1985 of a fair number of the most excellent graduates from science and engineering faculties (undergraduate level) finding employment in service industries, such as banking or insurance, not in manufacturing industries. This report also analyzed the reasons for and the prospects of this trend.

The second report was on the *Choice of Fields of Study Among University Applicants: How Many Young People in Japan are Planning to Study Science and Engineering in Universities?* (NISTEP Report No. 12, August 1990). Based on the results of a survey that sampled approximately 4,000 Japanese high school students, this study analyzed the students' attitude toward S&T and the choice of fields these students desired to enter into at universities.

The third report was on *How the Information on Science and Technology Activities Should Be Sent to Younger Generations—Based on the Analysis of the High School Students' Attitudes Toward Career Selection and Science and Technology* (NISTEP Report No. 24, October 1992). This report discussed the possibility of enhancing scientific and technological human resources through improving the scientific and technological information to Japan's younger generation. The contents of this report were based on the second analysis of survey results conducted for NISTEP Report No. 12. (See Figure 8.)

The fourth, a seminar report, described the *Savages in a Civilized Society—Young Peoples' Drift Away from Science and Technology* (NISTEP Seminar Report No. 26). This report is a record of a March 1991 speech by Mr. Shin'ichi Kobayashi, a visiting researcher of NISTEP. Mr. Kobayashi proposed a hypothesis based on *The Revolt of the Masses* by Jose Ortega y Gasset, a Spanish philosopher, and applied his hypothesis to the results of the above survey of high school students. (See Figures 9, 10, and 11.)

## Research on Science and Engineering Doctorates in Japan

Research in this group aims at strengthening policies to encourage doctoral programs in the S&T fields. It also covers the basic problems in doctoral courses.

The first study was a report on the *Quantitative Comparison of Science and Engineering Doctorates in Japan and the United States—Training of Researchers in Japanese Doctoral Programs* (NISTEP Report No. 7, December 1989). This study focused on the quantitative comparison of doctoral courses in science and engineering fields between Japan and the United States, especially the awarded number of doctorates between these two nations. (See Figure 12.)

The second report focused on *Conditions of the Strengthening of Doctoral Course in Natural Science Fields* (NISTEP Research Material No. 26, November 1992). Based on survey results and interviews with university professors, private enterprise leaders, researchers, and students in graduate courses, this study recommended some schemes to encourage Japanese scientific and technological research by strengthening doctoral courses.

## The Japanese S&T Indicator System

NISTEP published its first report on the Japanese S&T indicators system in September 1991. In this report, indicators concerning S&T were composed in Chapter 2 and in some clauses in other chapters.

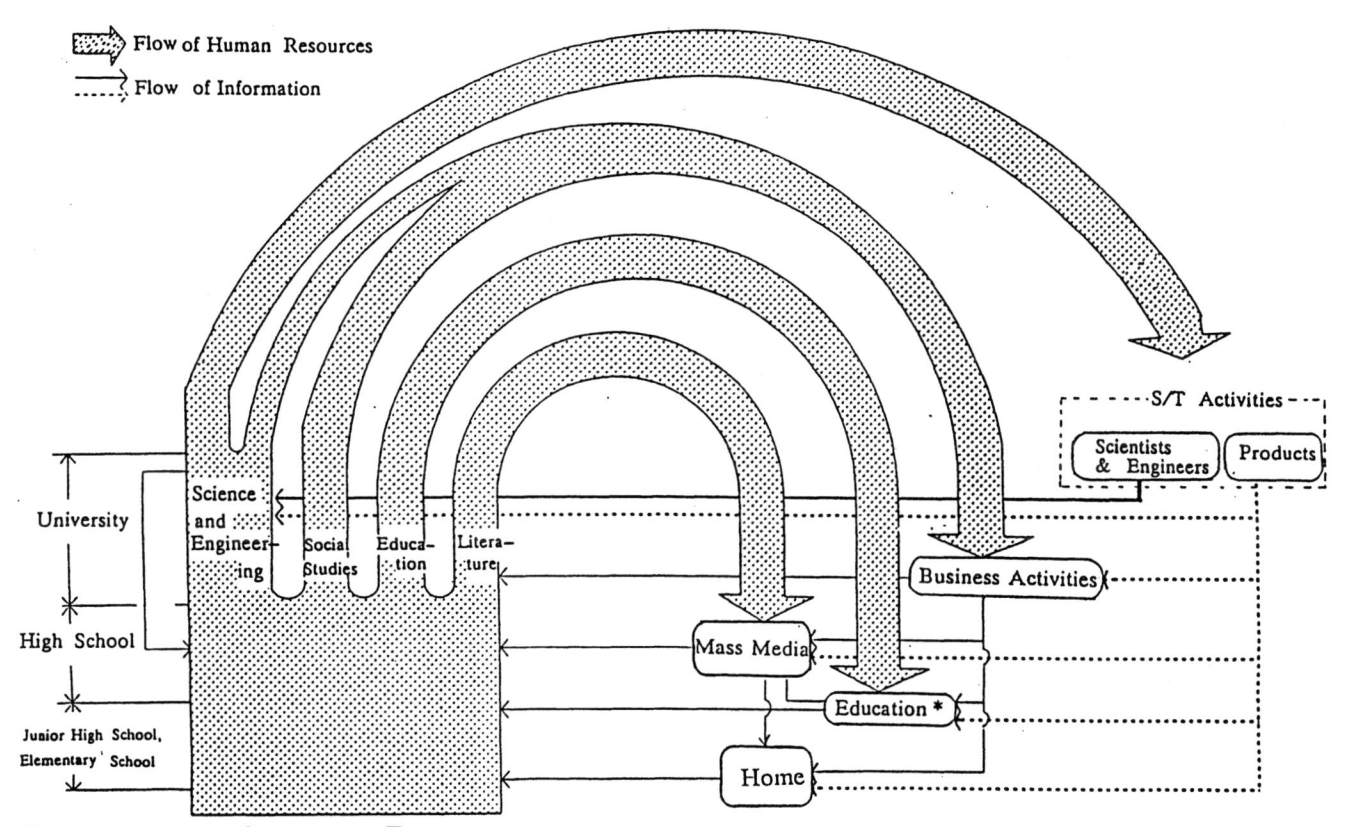

* ELEMENTARY AND SECONDARY EDUCATION
SOURCE: NISTEP Report No. 24, How the Information on Science and Technology Activities Should Be Sent to Younger Generations—Based on the Analysis of the High School Students' Attitudes Toward Career Selection and Science and Technology.

**FIGURE 8** Model of information flow on youth's career selection and science and technology activities.

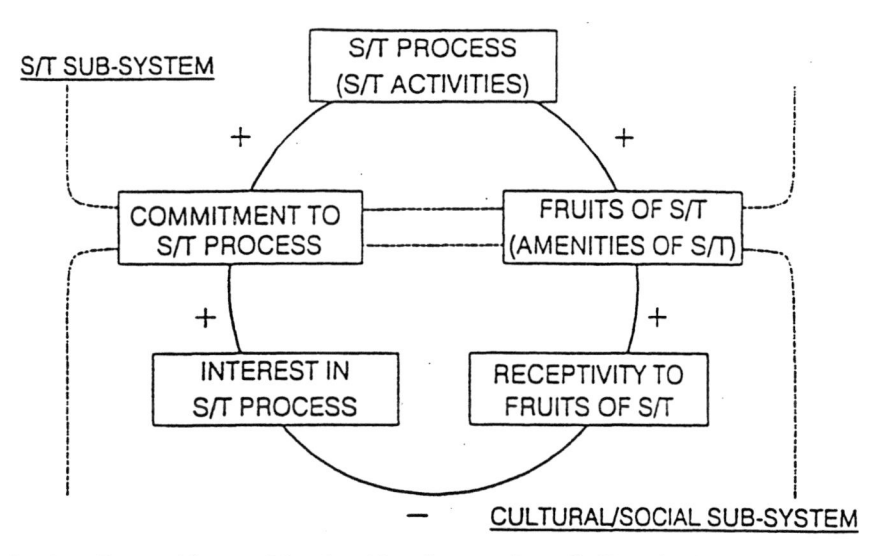

SOURCE: NISTEP Seminar Report No. 26, May 8, 1991, Savages in a Civilized Society—Young People's Drift Away from Science and Technology.

**FIGURE 9** Formulating the "Savages in a Civilized Society" hypothesis.

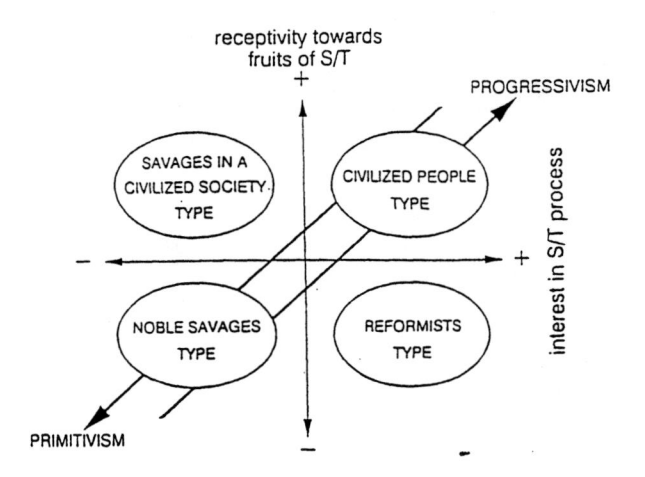

SOURCE: NISTEP Seminar Report No. 26, May 8, 1991, Savages in a Civilized Society—Young People's Drift Away from Science and Technology.

**FIGURE 10** Classification of people's attitudes toward science and technology.

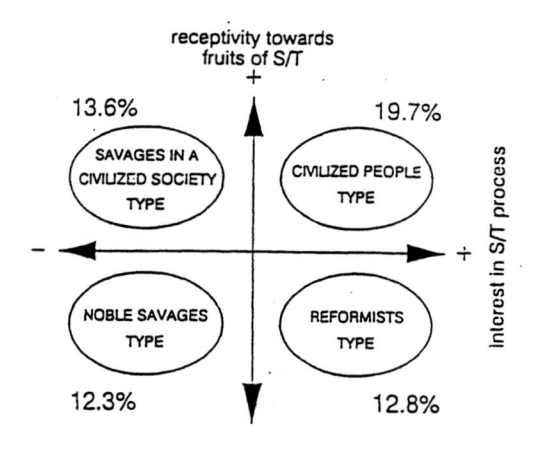

SOURCE: NISTEP Seminar Report No. 26, May 8, 1991, Savages in a Civilized Society—Young People's Drift Away from Science and Technology.

**FIGURE 11** Distribution according to survey of high school students.

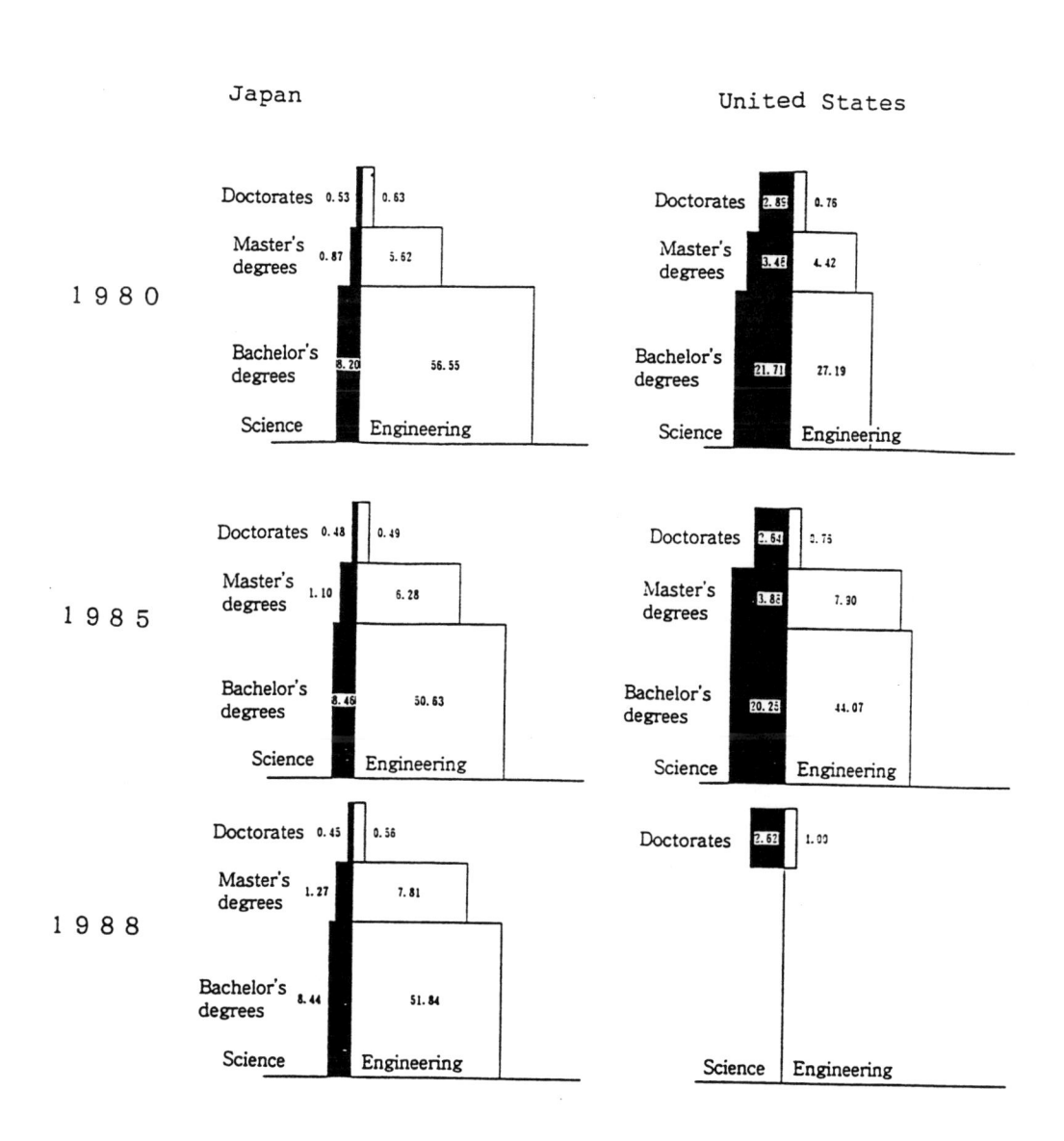

NOTE: Numerical values represent number of persons per unit population (100,000 people). Foreign students are not included. The number of Japanese doctorates per unit population is equivalent to the number of young doctorates, with 1980 representing fiscal 1979, 1985 representing fiscal 1984, and 1988 representing fiscal 1987.

SOURCE: NISTEP Report No. 7, Quantitative Comparison of Science and Engineering Doctorates in Japan and the United States.

**FIGURE 12** Comparison of science and engineering human resource supply structure in Japan and the United States.

## CONCLUSIONS

Through the above research, NISTEP has focused on three points about the essence of S&T career problems.

1. Encourage good relations in communication between the public (especially the younger generation), scientists and engineers, and organizations (government, private enterprises, etc.) regarding common S&T information.

2. Improve the educational system, especially in doctoral programs that train scientific and technological researchers in Japan.

3. Ensure high salaries for and good treatment of scientists and engineers in society.

In general, the Japanese people have created a highly industrialized society through great effort to modernize, westernize, and industrialize their nation over the past 100 years. During this period, Japan's social system and industrial structure has changed entirely, especially in the research fields of S&T. Although the gap between Japan and the western world continues to narrow, there are a few fields that are exceptions. These exceptions include the scarcity of doctoral programs in universities and the weakness of an informal public relations system in S&T information (especially in the academic world and the public).

Finally, I would like to emphasize that NISTEP aims to encourage international cooperation and to contribute to S&T policy research. The S&T career problem is one of the most important themes that NISTEP is now tackling.

# NOTES

1. The following chart describes the organization of NISTEP in 1993:

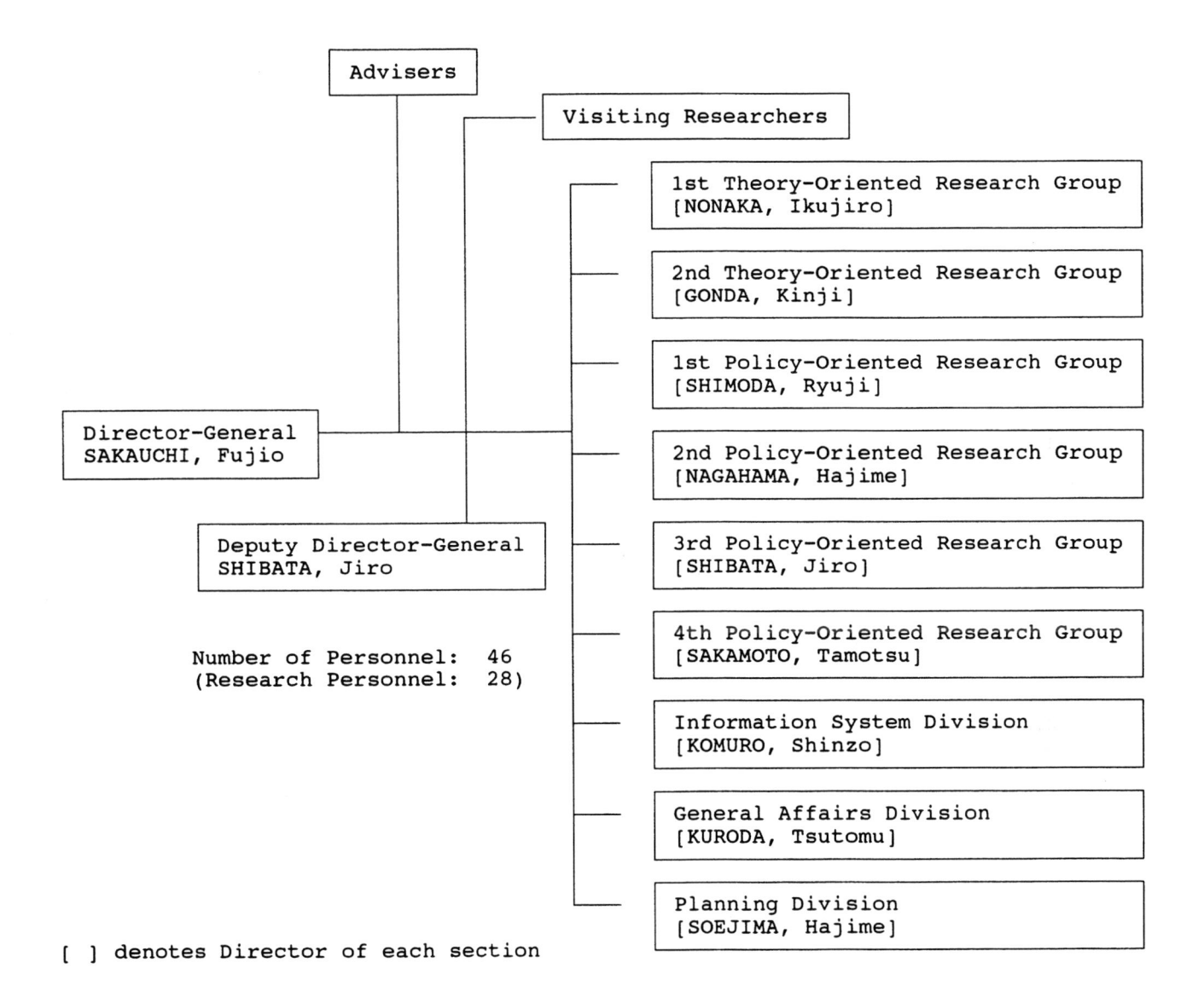

```
                                    ┌─────────────────────┐
                    ┌──────────────│     Advisers        │
                    │               └─────────────────────┘
                    │         ┌─────────────────────────┐
                    │    ┌───│   Visiting Researchers   │
                    │    │    └─────────────────────────┘
                    │    │    ┌──────────────────────────────────────┐
                    │    │ ──│ 1st Theory-Oriented Research Group    │
                    │    │    │ [NONAKA, Ikujiro]                    │
                    │    │    └──────────────────────────────────────┘
                    │    │    ┌──────────────────────────────────────┐
                    │    │ ──│ 2nd Theory-Oriented Research Group    │
                    │    │    │ [GONDA, Kinji]                       │
                    │    │    └──────────────────────────────────────┘
                    │    │    ┌──────────────────────────────────────┐
                    │    │ ──│ 1st Policy-Oriented Research Group    │
                    │    │    │ [SHIMODA, Ryuji]                     │
┌──────────────────┐│    │    └──────────────────────────────────────┘
│ Director-General ├┼────┤    ┌──────────────────────────────────────┐
│ SAKAUCHI, Fujio  ││    │ ──│ 2nd Policy-Oriented Research Group    │
└──────────────────┘│    │    │ [NAGAHAMA, Hajime]                   │
                    │    │    └──────────────────────────────────────┘
┌──────────────────────┐ │    ┌──────────────────────────────────────┐
│ Deputy Director-General├─┤ ──│ 3rd Policy-Oriented Research Group   │
│ SHIBATA, Jiro         │ │    │ [SHIBATA, Jiro]                     │
└──────────────────────┘ │    └──────────────────────────────────────┘
                         │    ┌──────────────────────────────────────┐
                         │ ──│ 4th Policy-Oriented Research Group    │
                         │    │ [SAKAMOTO, Tamotsu]                  │
Number of Personnel: 46  │    └──────────────────────────────────────┘
(Research Personnel: 28) │    ┌──────────────────────────────────────┐
                         │ ──│ Information System Division           │
                         │    │ [KOMURO, Shinzo]                     │
                         │    └──────────────────────────────────────┘
                         │    ┌──────────────────────────────────────┐
                         │ ──│ General Affairs Division              │
                         │    │ [KURODA, Tsutomu]                    │
                         │    └──────────────────────────────────────┘
                         │    ┌──────────────────────────────────────┐
                         └──│ Planning Division                      │
                              │ [SOEJIMA, Hajime]                    │
                              └──────────────────────────────────────┘
```

[ ] denotes Director of each section

2. NISTEP Report No. 17 includes the following S&T communications model:

## Science and Technology Communication Center
## Science Communication Center

(1) To promote intellectual activities and to provide needed incentives.

(2) To provide for adults and children access to the essential experience of scientific activities. In order for the public to become familiar with S&T, it is necessary to provide opportunities for them to come in contact with scientific and technological achievement.

(3) Society requires a system that nurtures the creativity as well as educates students about past scientific achievements. We have to construct a dual learning system that has two types of courses: one for static knowledge stock and the other for active inquiry.

(4) It is necessary for the public to provide a place where people can communicate with others, especially scientists and engineers. In the post-industrial society, social development will be guaranteed by full communication—consistent flow of information—with producers and consumers.

(5) To nurture abilities to discover and recognize invisible values or mechanisms. Namely, each person will be required to possess more ability in order to understand natural or social phenomena and movements (such as relativity, cooperation, contradiction, and circulation, etc.).

(6) Future development and public acceptance of S&T depend on the public's total recognition of S&T (e.g., goodness or badness of S&T, convenience or inconvenience of S&T, or what presumptions are presupposed and what limitations those technologies have, etc.).

(7) It is necessary for us to proceed and achieve the objectives outlined in the six items above. We recommend a "Science and Technology Communication Center" as a social infrastructure in which the public and scientists and engineers can participate together in scientific activities and events. Such centers would also function as learning centers (a lifelong education system) that would support school education viewpoints as part of a dual learning system.

3. The following material represents the Outline of the International Comparative Study of the Public Understanding on Science and Technology:

OUTLINE OF THE INTERNATIONAL COMPARATIVE STUDY
OF THE PUBLIC UNDERSTANDING ON SCIENCE AND TECHNOLOGY

(1.) Participation (as of October 1992): Japan, U.S.A., Canada, EC (12 countries), and China

(2.) Current surveys

| Country/Region | The Latest Survey | The Next Survey |
| --- | --- | --- |
| Japan | Nov. 1991 (Pilot survey: Jan. 1991) | Unsettled |
| U.S.A. | 1990 | Every two years |
| Canada | 1988 | 1992 (For youth: 1993) |
| EC | 1989 | 1992 |
| U.K. | 1988 | 1992 |
| China | 1991 | Unsettled |

(3.) International Council Meeting (every year)

| | | | | | | | |
| --- | --- | --- | --- | --- | --- | --- | --- |
| 1st | April | 1990 | London | 4th | | 1993 | Chicago |
| 2nd | Feb. | 1991 | Washington | 5th | | 1994 | London |
| 3rd | Oct. | 1992 | Tokyo | 6th | | 1995 | Beijing (prospect) |

(4.) Liaison Centers

- Japan: 2nd Policy-Oriented Research Group, National Institute of Science and Technology Policy
- USA: Science and Engineering Indicators Program, National Science Foundation and the ICASL, Chicago Academy of Sciences
- EC: EC Commission DG XII (Brussels) and the Science Museum Library (London)

# REFERENCES

2nd Policy-Oriented Research Group. 1992. The 5th Technology Forecast Survey - Future Technology in Japan (Japanese and English).

2nd Theory-Oriented Research Group. 1989. Preliminary Study on Regional Promotion of Science and Technology Throughout Japan (Interim Report) (Japanese).

4th Policy-Oriented Research Group. 1990. Development and Use of Biotechnology and Its Influence: Problem on Practical Use of Biotechnology (Japanese).

4th Policy-Oriented Research Group. 1989. Investigation of the Situation and the Structure of Energy Consumption in Asia and the Global Problems Anticipated Due to Increasing Energy Utilization (Interim Report) (Japanese and English) (abstract only).

4th Policy-Oriented Research Group. 1990. Preliminary Study on Promotion of Regional Science and Technology Throughout Japan (Japanese).

Hirano, Y. and C. Nishigata. 1990. "Basic Research" in Major Companies of Japan (Japanese and English).

Kagita, Y. and F. Kodama. 1991. Ratio Analysis of R&D Expenditure vs. Capital Investment in Japanese Manufacturing Companies: From Producing to Thinking Organizations (Japanese and English).

Kato, N., Y. Ogawa, T. Koike, T. Sakamoto, and S. Sakamoto, et.al. 1991. Analysis of Structure of Energy Consumption and Dynamics of Emission of Atmospheric Species Related to the Global Environmental Change (SOx, NOx, CO2) in Asia (Japanese and English).

Kiba, T. and F. Kodama. 1991. One Approach on Measurement and Analysis of International Technology Transfer: A Case Study on Japanese Firms' Direct Investment into East Asian Countries (Japanese and English).

Kikuchi, J., S. Mori, Y. Baba, and Y. Morine. 1990. Development of Input-Output Model for Science, Technology and R&D (Interim Report): R&D Dynamics (Japanese and English).

Kikuchi, J., S. Mori, and M. Morino. 1989. Development of Input-Output Table for Science, Technology and R&D (Interim Report): Theoretical Framework of Input-Output Model for Science, Technology and R&D (Japanese).

Koba, K., S. Ishiyama, and H. Nagahama. 1992. Background and Professional Activities of Natural Science Research Personnel in Japan (Japanese).

Kobayashi, S., H. Endo, E. Sato, and Y. Hirano. 1992. How the Information on Science and Technology Activities Should Be Sent to Younger Generations—Based on the Analysis of the High School Students' Attitudes Toward Career Selection and Science and Technology (Japanese and English).

Kobayashi, S. 1991. Savages in a Civilized Society—Young People's Drift Away from Science and Technology (Japanese and English).

Mori, S., J. Kikuchi, Y. Baba, and H. Mitsuma. 1992. Development of Input-Output Table for Science, Technology and R&D: Some Results and Policy Implications (Japanese).

Muto, E. and Y. Hirano. 1991. Government Laboratories and Basic Research: Toward the Promotion of Basic Research in Government Laboratories (Japanese and English).

Nagahama H., T. Kuwahara, and T. Nakahara. 1989. Public Attitudes on Science and Engineering: Based on Opinion Survey Results (Japanese and English).

Nagahama, H., T. Kuwahara, A. Nishimoto, and the STSC Study Group. 1991. Science, Technology, Society and Communication (Japanese and English).

Nanahara, T., and F. Niwa. 1990. Trends of R&D Activities in Japanese Companies Using Patent Statistics (Japanese).

Nishigata, C., A. Nakanishi, and Y. Hirano. 1989. Employment Trends of Science and Engineering Graduates (Japanese and English).

Nishigata, C. and Y. Hirano. 1992. Increasing the Number of High Quality Science and Engineering Taught - Course Doctorates in Japan (Japanese).

Nishigata, C. and Y. Hirano. 1989. Quantitative Comparison of Science and Engineering Doctorates in Japan and the United States: Training of Researchers in Japanese Doctorate Courses (Japanese and English).

Nishimoto, A. and H. Nagahama. 1991. The Interchange of Researchers and Engineers Between Japan and Other Countries: Study Based on Annual Report of Statistics on Legal Migrants, Statistics on Foreigners Staying in Japan, and Annual Report of Statistics on Japanese Nationals Overseas (Japanese and English).

Nishimoto, A. and H. Nagahama. 1990. Trends of Science and Technology Activities in Japan Using Science and Technology Awards Statistics: Persons of Scientific and Technological Merits, Commendation by the Minister of States for Science and Technology (Japanese).

Niwa, F., H. Tomizawa, F. Hirahara, F. Kakizaki, and O. Camargo. 1991. The Japanese Science and Technology Indicator System: Analysis of Science and Technology Activities (Japanese and English).

Sato, E., H. Kikuchi, and Y. Hirano. 1990. Choice of University Applicants Among Fields of Study: How Many Young People in Japan are Planning to Study Science and Engineering in Universities? (Japanese and English)

Shirai, I. and F. Kodama. 1989. Quantitative Analysis on Structure of Collective R&D Programs by Private Corporations in Japan (Interim Report) (Japanese and English).

Watatani, H., T. Yamamoto, K. Gonda, and T. Sakamoto. 1992. Study of Regional Science and Technology Promotion: Analysis of Science and Technology Policies by Local Governments (Japanese).

# Developing Data Systems on Trends in Science and Technology Careers

## Glynis Breakwell

My task is to explore how the research described in the papers of Westholm and Nagahama assist us in monitoring trends in science and technology (S&T) careers. It is also my task to examine what other factors should be taken into account when developing information collection procedures that will provide adequate data systems for policymaking concerning human capital resources in S&T.

### WHAT DATA ARE NEEDED?

Three types of data are actually needed in policymaking in this area:

Type 1.  Data on what S&T personnel exist at any one point in time.

Type 2.  Data on the patterns of change in S&T personnel over time.

Type 3.  Data on the factors that influence changes in both the quality and quantity of S&T personnel.

Type 1 data are what Westholm calls information on the stock of S&T personnel. Type 2 data are what he calls the flow of such personnel. Type 3 data, which are by far the most complex, seem to me to be the main target for the substantive work described by Nagahama. It could be argued that a fourth type of data should also be collected: on availability of jobs per se. The availability of jobs should not be confused with availability of careers. Data on careers would entail a longitudinal perspective following individuals over time.

### WHAT PROBLEMS MUST BE OVERCOME IN CREATING DATA SYSTEMS?

#### Problem 1: Comparability—Geographical and Temporal

Any decent data system requires unambiguous categories or units of measurement. We are certainly not faced with a dearth of stock and flow data. As Gunnar Westholm explained, these exist for extended time periods in some countries. We lack comparable data—comparable, that is, across time and across place.

At one level, the lack of comparability is merely a product of confusion of terminology. Either the same word is given different usages or different words are given the same usage. There is, for example, the perennial problem of the meaning of technician. At a rather more significant level, comparability is lost because data collection procedures differ. In this case, the category or unit of measurement is agreed, but the data collected varies in accuracy or rigor. The reasons for differences in data collection procedures are numerous. They should be examined by all who are interested in ensuring valid and reliable data systems. Only by identifying the nature of these differences and the reasons why they are perpetuated will we begin to know how to regulate the research community in such a way as to achieve data comparability.

There is, however, a superordinate question: Why do we need international comparability? This would imply either the existence of international policymaking or an international S&T job market. Neither currently exist except for a relatively small number of high-level jobs.

## Problem 2: Specificity

Achieving data at sufficiently specific levels of description has been a problem. Aggregation of data at the level of the lowest common denominator has been necessitated by the low levels of comparability. As categories or units of measurement have become more specific, their implications have been more selective and idiosyncratic. Westholm rightly emphasizes OECD's problems in this regard. He argues they result in macro rather than micro levels of data being used in any comparisons across countries. There is obviously some trade-off between comparability and specificity at the moment. This is not inevitable, but if it is to be avoided, there will need to be even greater international collaboration in establishing the definition of the categories or units of measurement used.

There is, however, another problem that militates against the development of more specific S&T data. It stems from the nature of S&T activities and the way in which they are embedded in economic and policy systems. At the heart of S&T development is the creation of new job categories and different career structures. This makes it necessary for data systems to anticipate trends in order to ensure that the information collected is relevant and up-to-date. Such anticipation is not often possible. The more specific the unit of measurement, the more it is open to becoming redundant as scientific and technological changes transform labor demands. It seems that it is important for policymakers to decide the levels of specificity that will optimize planning. These levels will, of course, vary according to the precise S&T domain under consideration. It is conceivable that in areas where changes are less frequent or significant, high levels of specificity will be feasible. The point is, however, that the level should be dictated by planning requirements rather than convenience. It would have been useful if the papers by Westholm and Nagahama had considered what new structures might be needed at the international governmental level to establish optimum specificity in data sets.

## Problem 3: Periodicity

Regular periodic data collection is needed but does not actually happen in most countries. Currently, there is an unhealthy reliance upon ad hoc studies or, as Westholm argues, tacking extra questions on to other surveys. What seems to be needed is the education of national data collection agencies about the purpose and value of having, firstly, regular data and, secondly, internationally coordinated schedules of data collection. National agencies need to be persuaded that involving the international research community in developing the instruments used in data collection is worthwhile. It is necessary for us to prove the policy value of such statistics. Examples of the proven worth of these types of data sets would have been valuable in the papers by Westholm and Nagahama.

If systems of regular periodic data collection were established, they might act as something of an antidote to one of the major problems in existing practice. Typically there are substantial time lags between data collection, analysis, interpretation, publication, and policy application. This often means that policy is based upon statistics that are out-of-date and misleading. Regular data sweeps would encourage streamlining of data handling as routines were introduced. Regularity would allow better trend analysis and once established these trends would focus the direction of initial analyses, speeding interpretation. Needless to say, regular sweeps would also be likely to create a group of researchers with specific expertise.

## Problem 4: Structuring the Data

Issues of comparability, specificity, and periodicity tend to distract us from the equally significant problems associated with determining what data are really necessary. Westholm's model of stocks and flows of S&T personnel is undoubtedly a useful starting point in examining these problems. It is underpinned by the UNESCO distinction between the total stock of qualified manpower and the number of economically active qualified manpower (which begs the question: does qualified refer to educational or occupational achievements? But that can be left aside for now). This distinction has to be further broken down into those who are qualified and working in S&T and those who are qualified but working in some other area. These two criteria for dichotomizing the stock of S&T

personnel essentially yield the model proposed. I cannot accept the label Westholm gives the flow diagram; it is a way of systematizing what we need to know, but it is not a theory. It is actually the cross-classification, which he says is too ambitious, of inflow and outflow information with detailed breakdowns by, for example, type of personnel, occupation, sector of employment, educational background, age, gender, nationality, etc., which would allow an explanatory or predictive theory to be developed. Such cross-tabulation is the only way to generate a model specifying what factors influence the patterns of inflow and outflow.

The schematic, as it stands, focuses solely upon the extent to which the active stock of S&T personnel (STP) is maintained. Another complementary schematic is needed to represent movement within the active stock between categories or career or job. Such a schematic would need to represent changes in the structure of the STP stock in terms of age, gender, sector of employment, fields of qualification, location, etc., across time. These changes in the structure of the stock are important in policymaking.

Once the changes in the structure of the STP stock are treated seriously, another question is raised. The topic of this conference is trends in S&T careers, not jobs. The concept of career implies the need to consider what happens to a person over time in terms of education and employment. The model of data collection implicit in Westholm's schematic is cross-sectional, not longitudinal. It would make it difficult to establish anything about trends in scientific and technical careers except through tenuous extrapolation. A simple example serves to illustrate this point. Let's say that half of the STP stock occupied research and development jobs and the other half did not (I realize this is unimaginable) at time one (1993) and the same proportions exactly were found to apply at time two (2093). (It could be that the individuals in each sector had remained the same or it could be that some proportion of individuals had switched sectors. As long as the proportions switching were identical, the results would be the same.) From such data, we know nothing about careers per se. In order to understand something about careers, we need information on one individual across his or her life span. This is, of course, the area where Nagahama's report offers great hope. It is evident, even from the very brief descriptions given, that the Japanese research is focusing upon the structure of scientific careers.

Nagahama's summary of research activity at NISTEP reveals an emphasis on modeling the factors that influence changes in the inflow and outflow of STP. The studies of public understanding of science, the assessment of the impact of educational policy, and the effectiveness of intervention techniques are all motivated by the need to understand how people come to choose and maintain effective careers in S&T. Of course, such studies are based on quite different types of information from the stock and flow model of Westholm. These data are probably even less open to tight international comparability. In fact, comparability in such studies at the level of the operational definition of indexes might be pointless. Cultural differences in the factors affecting career decisions may make cross-cultural comparability in data collection redundant as well as impossible. Cross-cultural comparability will probably have to be achieved at the level of theoretical rather than operational constructs. This may not be too severe a problem if the forecasts within cultures are good. After all, these data are designed to serve a different function from the statistics on overall stocks and flows. While stocks and flows data primarily reveal availability, these data are the basis for manipulating availability.

The ultimate objective of such modeling of the factors influencing career choice and change would be prediction and manipulation of trends in STP stocks. Westholm regards prediction or forecasting as hazardous and too difficult. Yet I am glad to see Nagahama's Institute attempting it at some levels. Without forecasting, such exercises become merely descriptive—a sort of statistical historicism. Since the data are always retrospective, they come to have a less influential role in policymaking.

## CONCLUSION:
### Confronting The Problems

Specification of the problems takes us some way toward their solution. In fact, it seems we know pretty well what sorts of data are needed on stocks and flows of STP. We also know what factors should be explored in establishing predictive models. The difficulties lie at the pragmatic rather than the conceptual level. It is the financial and political constraints that make progress slow. We should be discussing how to overcome them.

# PART II

# Analyzing Trends in Science and Technology Careers: The Longitudinal Approach

# Summary of Introductory Comments

## Paul Baltes

There are several ways to think about the development of science and technology (S&T). For instance, we can think about the development of individuals across their life span; that is, in terms of achieving their full potential to contribute to the S&T process and in terms of their active participation in the S&T workforce. We can talk about the development of institutional structures and the interactions between people and the S&T system. We might focus on generational dynamics such as the transformation of social roles and the age at which society expects someone to step into the role of a scientist or engineer. Finally, there is the concept of the life cycle of research programs. In all of these scenarios of S&T development, there is an implicit interest in monitoring both change and constancy. What differs between these scenarios is the unit of analysis and the nature of the time continuum.

Each of these vistas on the development of S&T, and especially their interrelationships, requires longitudinal, repeated-measurement methodology in one form or another (Magnusson, 1993; Nesselroade & Baltes, 1979); that is, research that repeatedly observes the same unit (individuals, groups, institutions) over time. However, largely for reasons of economy and investigator time, the dominant methodological tendency has been to conduct cross-sectional research to monitor development—taking, for example, one snapshot of individuals of different ages, of a labor force, or of an educational cohort. Such cross-sectional studies—studies conducted at one point—make it virtually impossible to disentangle the many factors that shape the time course of an individual, a program, or an institution. Moreover, the outcome of cross-sectional, one-time comparisons confound individual (age) change with historical change. For example, when comparing 40- and 60-year-olds in 1990, these two groups differ not only in age, but also in the historical time during which they grew older (Schaie, 1965). Thus, strictly speaking, cross-sectional studies fail to provide much insight into the underlying life course mechanisms of development, whether they are personal, institutional, or societal.

I emphasize this point also because, historically, scientific insights into understanding the limitations of cross-sectional studies in the analysis of change owes much to Belgium. We are gathered here at a place that possesses characteristics of a true genius loci, where cohort and age-based methodology are concerned. In the nineteenth century, the Belgian Adolphe Quetelet (1842) made an important advance in the application of statistical analysis to social data: he formulated the concept of the "average man." Particularly relevant to today's conference is his methodological argument for longitudinal research and the need to repeat "age studies" over time and cohorts to avoid the "disturbances" created by historical period effects. As a result of Quetelet's work, we mark the nineteenth century as the beginning of research on the developing individual in a changing society and the study of age-cohort relationships.

We have learned a lot about the concept of human development in the years following Quetelet's seminal work. The study of human development has evolved and the concept of life span analysis and associated life course methodology (such as cohort studies) have gained favor (Baltes, 1987; Featherman, 1983). Two points deserve highlighting from this century-long quest for adequate methodology and adequate data:

- There are no methodological shortcuts in understanding the life course of individuals and educational institutions. If the focus is on change, cross-sectional studies are but a first step.

- There is an important role for longitudinal research in disentangling the complex interactions of individuals and the environment in which they develop. Moreover, longitudinal studies of single birth cohorts need to be enriched by cohort-sequential studies.

In summary, for life span and life course researchers, cross-sectional, one-time data are insufficient. Rather, life span development and life course researchers argue that cohort-sequential and longitudinal data are critical for developing the knowledge bases required to understand human development in a changing society. Such methodologies can also be used to formulate effective policies for attracting individuals to S&T and for facilitating their transition into productive careers. The presentations that follow are illustrations of the need for longitudinal research and the policy implications they suggest.

# REFERENCES

Baltes, P. B. 1987. Theoretical Propositions of Life Span Developmental Psychology: On the Dynamics Between Growth and Decline. Developmental Psychology. 23:611-626.

Featherman, D. L. 1983. The Life Span Perspective in Social Science Research. Life Span Development and Behavior, P. B. Baltes and O. G. Brim, Jr., (eds) 5:1-59. New York: Academic Press.

Magnusson, D. 1993. Human Ontogeny: A Longitudinal Perspective. Longitudinal Research on Individual Development. 1-25. D. Magnusson and P. Casaer (eds). New York: Cambridge University Press.

Nesselroade, J.R. and P.B. Baltes (eds). 1979. Longitudinal Research in the Study of Behavior and Development. New York: Academic Press.

Quetelet, A. 1842. A Treatise on Man and the Development of His Faculties. Edinburgh: Chambers.

Schaie, K. W. 1965. A General Model for the Study of Developmental Problems. Psychological Bulletin. 64:92-107.

# A Demographic Approach to Studying the Process of Becoming a Scientist/Engineer

## Yu Xie

In this paper, I redefine career process as the collective experience of a birth cohort and propose a new demographic approach to studying the developmental process of becoming a scientist/engineer by following a synthetic birth cohort through its formative years of career development. The approach is dynamic rather than static, in the sense that it traces career changes over the life course of a cohort. At any given age, cohort members are identified as belonging to one of several states relevant to a scientific/engineering career. With data from longitudinal surveys, probabilities of cohort members' movements into and out of the different states are calculated as functions of age and population characteristics.

From these transitional probabilities, the process of becoming a scientist/engineer is modeled assuming a time-inhomogeneous Markov process commonly seen in standard multistate life tables. The time-inhomogeneity property of the Markov model makes an analysis adopting this approach non-parametric, descriptive, and capable of reproducing observed statistics with good data. Cross-sectional statistics on scientific careers can be generated from such a demographic model, for the size of any population is ultimately determined by inflow rates into and outflow rates out of the population.

The proposed approach has two important advantages over cross-sectional studies, which can only yield static snapshots of the population size with no information on the dynamic process of inflow and outflow. First, it makes it feasible to infer current and future descriptive statistics and other useful information on science/engineering (S/E) careers for any population or subpopulation with observed or hypothesized transition probabilities. Second, it allows the researcher to locate sources of attrition, especially for women and disadvantaged minorities, along the pipeline to becoming a scientist/engineer.

One major constraint for implementing the proposed approach in practical settings is the lack of longitudinal data, which are required for such dynamic analyses. When data from a true longitudinal design are unavailable, I propose that a synthetic cohort be constructed from different sources. For a demonstration of the new approach, I piece together data from two large data sets representing U.S. youth as they grow up between ages 13 and 32. The 1987-1991 Longitudinal Study of American Youth (LSAY) is used to obtain middle and high school students' (grades 7 through 12) interests in science education and changes of their interests over time. The 1972-1986 National Longitudinal Study of the Class of 1972 (NLS-72) is used to obtain information on years beyond high school, i.e., youth's probabilities of majoring in science, receiving science degrees at bachelor's and master's levels, and working in scientific occupations. Men and women are analyzed separately.

## BASIC CONCEPTS AND METHODS

### Cohort

In a classic article, Ryder (1965) defines a *cohort* as "the aggregate of individuals (within some population definition) who experienced the same event within the same time interval" (p.845). For example, all individuals born at the same time (say within a given calendar year) make up a birth cohort. Similarly, events such as marriage and school entry also define marriage and school cohorts. In this paper, I assume the equivalence between birth cohorts and school cohorts up to secondary education. This is a reasonable assumption because most children start school at about the same age and progress through elementary and secondary education on a similar schedule. For students attending postsecondary schools, I pay closer attention to different schooling paths so that a person currently not in school may enter school at a later date.

Ideally, we would like to observe all career changes of a real cohort for its entire work-relevant history, from childhood to retirement. This would allow us to accurately model the life course career process of the cohort. Such longitudinal designs, however, are unrealistic in practice not only because they are too expensive, but also because they take too long and thus cannot yield even tentative answers to important questions currently faced by today's social scientists.

One common solution to this dilemma, often adopted by demographers in studies of fertility and mortality, is to construct age-specific vital rates from a cross section and assume them to be experienced by a hypothetical cohort. For instance, the total fertility rate is the expected total number of children a woman would have if she followed the entire age-specific fertility schedule of a given period, and life expectancy is the expected total number of years a newborn child would live if he or she were subject to the age-specific mortality schedule of a given period.

An excellent application of this approach to the study of scientific personnel is found in Berryman (1983), who compiled detailed cross-sectional statistics at salient points in the educational pipeline (such as degrees at all levels) by race and gender. One major drawback of this solution, however, is its inability to uncover dynamic processes underlying cross-sectional data. For example, Berryman was unable to examine the changes in enrollment status and field of study and their variations across gender and race, even though she clearly realized the importance of such transitions.

Limited longitudinal studies, a middle ground between purely cross-sectional designs and ideal longitudinal designs, have gained more popularity and acceptance in recent years. By limited longitudinal studies, I mean that researchers follow a group of subjects only for a limited duration. Examples are the NLS-72, LSAY, High School and Beyond (HS&B), National Educational Longitudinal Survey (NELS), and the Survey of 1982-1989 Natural and Social Scientists and Engineers (SSE).

Limited longitudinal studies could be cohort-based, such as NLS-72, LSAY, HS&B, and NELS, or population-based, such as SSE. While the sampling frames of NLS-72, LSAY, HS&B, and NELS were school cohorts, the sampling frame of SSE was the population of scientists identified by the 1980 U.S. census. Because there are currently many large, nationally representative, cohort-based limited longitudinal studies available, I propose to piece together the experiences of different cohorts to form a synthetic cohort. Here I define a synthetic cohort as a hypothetical cohort whose life history is constructed from different real cohorts in a supplementary manner. Even though the synthetic cohort is not real, segments of the cohort's experiences are real. One major advantage to this approach is that we can observe cohort members' transitions into and out of different states. This enables researchers to study "social dynamics" (Tuma and Hannan, 1984), which is not possible with cross-sectional data. See Figure 1 for an illustration of a synthetic cohort (LSAY1 and LSAY2 respectively refer to the first and second cohorts in LSAY).

## Career Process Redefined

Traditionally, career process has been defined as an *individual's* career history.[1] This is true in (1) trait and factor theory, social learning theory, and developmental theory in psychology (e.g., Brown and Brooks, 1990), (2) status attainment models in sociology (e.g., Sewell, Haller, and Portes, 1969), and (3) rational choice theory and human capital theory in economics (e.g., Freeman, 1971). One difficulty with this conventional perspective is that an individual's career choices not only change frequently over the life course, but also change in such irregular sequences that they cannot be easily characterized by a unidirectional development model.

In this paper, I redefine *career process* as the collective experience of a birth cohort. While individuals of the cohort may change their career choice frequently and irregularly, the cohort as a whole may well exhibit regular patterns of career process that are

The experiences of different cohorts are pieced together in an analysis.

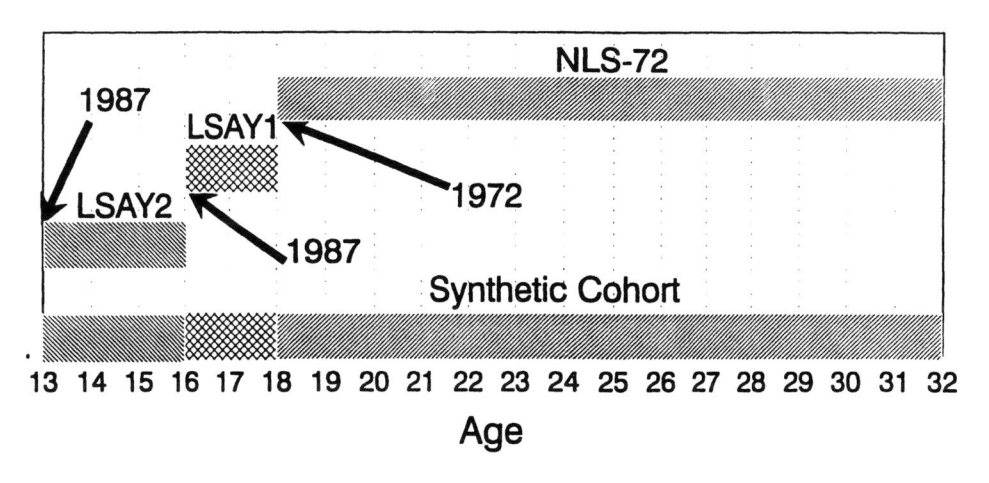

**FIGURE 1** A synthetic cohort approach.

subject to scientific scrutiny.[2] For a given cohort, the career process coincides with maturation and aging, as in the case of an individual. Different from the individual-based definition, however, the cohort-based definition of career process allows the researcher to characterize the process using aggregate statistics with a certain degree of accuracy. Take the process of becoming a scientist/engineer as an example. Over time, some members of a birth cohort may stay in, move into, or move out of the S/E pool. Transition probabilities for these movements and non-movements are interesting characteristics of the cohort. In this paper, I explain how to study a cohort-based career process using these probabilities.

### "Bathtub" Model of a Population

Any population can be described by the "bathtub" model: some people move in from outside while some people move out from inside. Simple as it appears, the "bathtub" model is actually a dynamic model. If the outflow rate exceeds the inflow rate for a sustained duration, the "bathtub" would eventually dry out; if the inflow rate exceeds the outflow rate for a sustained duration, the "bathtub" would eventually overflow.

Now consider the S/E pool as a population somehow-defined. For example, from early grades through high school, the population can be defined as students who are interested in science subjects and plan to attend college. In undergraduate and graduate years, the population can be defined as students majoring in science or obtaining science degrees. In the labor force, the population may be defined as workers in scientific occupations. As a birth cohort progresses through different stages, some cohort members move into the S/E pool while others move out of the pool. The task of a demographic analysis is to study the flows into and out of the pool and their implications for the process of becoming a scientist/engineer.

### STATES AND TRANSITIONS

A *state* (denoted as $s$) is a distinct, well-defined, temporarily stable condition. A set of mutually exclusive and exhaustive states constitute a *state space* (denoted as $S$, $s \in S$, where $S = 1, \ldots M$). Let $t$ denote time, which is assumed to be discrete in this paper (i.e., $t = 1, \ldots T$).

Transition probability, $p_{ij}(v, w)$, is the probability that an individual will be in state $j$ at $t=w$ given he or she is in state $i$ at $t=v$ ($v < w$):

$$p_{ij}(v,w) = Prob(s=j, t=w \mid s=i, t=v).$$

Let $P(v, w)$ be a square matrix that contains all elements of $p_{ij}(v, w)$:

$$P(v,w) = \begin{vmatrix} p_{11}(v,w) & \dots & p_{1m}(v,w) \\ & \dots & \\ p_{m1}(v,w) & \dots & p_{mm}(v,w) \end{vmatrix}$$

Likewise, the marginal (unconditional) distribution of states, $p_i(v)$, can be defined as

$$p_i(v) = Prob(s=i, t=v).$$

Let $\underline{p}(v)$ be a column vector that contains all elements of $p_i(v)$:

$$\underline{p}(v) = \begin{vmatrix} p_1(t=v) \\ \dots \\ p_m(t=v) \end{vmatrix}$$

From these definitions, it is easy to see the following relationship

(1) $\quad \underline{p}(w) = P(v,w)' \underline{p}(v),$

where $P(v, w)'$ is the transpose of $P(v, w)$. That is, the probability that a person is in the $j$th state at time $w$ is:

(2) $\quad p_j(w) = \sum_{i=1}^{m} [p_i(v) p_{ij}(v,w)]$
$= p_1(v) p_{1j}(v,w) + p_2(v) p_{2j}(v,w)$
$\dots p_m(v) p_{mj}(v,w)$

## MARKOV PROCESS

Markov processes are commonly assumed in multistate life tables (Namboodiri and Suchindran, 1987, Chapter 9). In this paper, I borrow this position and construct models based on the Markovian assumption.[3] That is, it is assumed that the state distribution at time $v+1$, $p_i(v+1)$, only depends on the state distribution at time $v$, $p_i(v)$, not on state distributions prior to $v$. Note that the Markovian assumption does not rule out past history prior to time $v$ as totally irrelevant to the state distribution at time $v+1$. It only states that past history prior to $v$ is relevant to $p_i(v+1)$ only insofar as it affects $p(v)$. In the language of structural equation models, prior history has no direct effects, but only indirect effects through $p_i(v)$. The Markovian assumption means that we can obtain the marginal (unconditional) distribution of states at time $v+1$ from $P(v, v+1)$ and $\underline{p}(v)$:

(3) $\quad \underline{p}(v+1) = P(v,v+1)' \underline{p}(v).$

Likewise,

$\underline{p}(v+2) = P(v+1, v+2)' \underline{p}(v+1)$
$= P(v+1, v+2)' P(v,v+1)' \underline{p}(v)$
$= P(v, v+2)' \underline{p}(v)$

(4)

$\underline{p}(v+3) = P(v+2, v+3)' \underline{p}(v+2)$
$= P(v+2, v+3)' P(v+1, v+2)' P(v, v+1)' \underline{p}(v)$
$= P(v, v+3)' \underline{p}(v)$

...

In general, the following chain rule is true:

(5) $\quad P(v,w) = P(v,v+1) \, P(v+1,v+2) \dots P(w-1,w).$

In other words, the transition probability matrix from time $v$ to time $w$ is the product of all transition probability matrices connecting time $v$ and time $w$. One implication of the property is that skipping intermediate steps in calculating transition matrices is not significantly detrimental if they are not of primary concern, so long as we obtain information about transitions before and after the intermediate steps.

It should be pointed out that the Markov chain model described by equations (1) through (5) follows the tradition of multistate life tables (Namboodiri and Suchindran, 1987) in that all transition matrices are time-dependent and subject to nonparametric estimation. This contrasts to the alternative treatment assuming time-homogeneity or a parametric form for transition matrices, a class of Markov models frequently discussed in the literature on stochastic processes (e.g., Bartholomew, 1973). In a sense, the Markov model of equations (1) and (5) does no more than decompose the observed marginal distribution of a later period in terms of transitions and initial conditions of earlier periods, if data for a true cohort are used. This methodology has been applied to studies of labor force participation in the form of working life tables (Hoem, 1977) and schooling in the form of school life tables (Land and Hough, 1989). This paper extends it to the study of the career process of becoming a scientist/engineer with the construction of a synthetic cohort.

## DATA

I combine data from two sources for the construction of a synthetic cohort between ages 13 and 32. The first source is the 1987-1991 LSAY, and the second is the 1972-1986 NLS-72. In LSAY two high school cohorts were followed up every semester, one from grade 7 in 1987 to grade 10 in 1990 and the other from grade 10 in 1987 to grade 12 in 1989. The NLS-72 cohort was grade 12 in 1972 and was followed up in 1973, 1974, 1976, 1979, and 1986.[4] I treat the three cohorts as if they were part of a single cohort. The surveys provide enough information to cover the hypothetical cohort continuously from ages 13 to 32.[5]

The pitfall of this research strategy is, of course, that the experiences of the synthetic cohort do not represent those of any real cohort. In this paper, the data sets used might be problematic, as the earlier years of the synthetic cohort were observed much later (starting in 1987) than the later years of the synthetic cohort (starting in 1972). Without better data, my analysis of the life course process of becoming a scientist/engineer assumes that career process is relatively stable, i.e., age-dependent rather than cohort- or period-dependent. This assumption is consistent with an observation made more than 40 years ago by Ginzberg and his associates (Ginzberg et al., 1951) that the career process is a developmental process, thus, an age-dependent process.

Each fall, the LSAY survey asked high school students whether they would "enroll in a four-year college or university" upon graduation. I consider those students who answered "yes" as intending to obtain postsecondary education. The same questionnaire also asked whether students agree or disagree with the statement "I enjoy science" as follows: "strongly agree," "agree," "not sure," "disagree," and "strongly disagree." I classify students who responded "strongly agree" *and* planned to enroll in a four-year college or university as belonging to the state of "intended science/engineering postsecondary education." This measure is constructed for all high school years of both cohorts in LSAY.

In each follow-up, NLS-72 respondents were asked to report their actual or intended fields of study if they were attending postsecondary school. The answers were coded into six-digit numerical codes. From the 1972 base year to the 1979 fourth follow-up, NLS-72 used a coding system described by *A Taxonomy of Instructional Programs in Higher Education*, commonly referred to as HEGIS (Huff and Chandler, 1970). For the fifth follow-up, NLS-72 changed to a new system described by *A Classification of Instructional Programs* (Malitz, 1981). From the two systems of instructional programs, I extracted a number of detailed codes as S/E fields of study.[6] In 1976 and 1979, NLS-72 ascertained whether respondents had obtained bachelor's and master's degrees. For those with positive responses, NLS-72 ascertained detailed fields of their degrees. In the 1986 fifth follow-up, NLS-72 changed the question and collected information regarding respondents' highest degrees (from baccalaureates to doctorates); data on fields of degrees, however, were not collected. To solve this problem, auxiliary information is used with additional assumptions. Respondents who undertook postsecondary education between 1979 and 1986 were asked to report their latest fields of study. Thus, for a respondent who obtained additional postsecondary education and whose highest degree in 1986 was more advanced than or equal to his or her highest degree in 1979, his or her latest field of study is imputed as the field of his or her highest degree.

I hereby create the following seven educational states:

(0) secondary education or college dropout
(1) intended or actual non-S/E postsecondary education
(2) intended or actual S/E postsecondary education
(3) non-S/E bachelor's degree
(4) S/E bachelor's degree
(5) non-S/E master's degree
(6) S/E master's degree

Figure 2 presents schematic flows among the seven educational states; the solid lines represent typical flows and the dotted lines represent untypical flows.[7]

From the above operational definitions of the states, I calculate, separately for males and females, 13 transition probability matrices for the hypothetical cohort, given in Tables A1 and A2. The reported rates and counts were appropriately weighted to reflect the sampling designs and non-responses so that resulting transition rates are the best estimates of their corresponding conditional probabilities in the population.[8] These transition rates are mainly used in analysis to be reported later. It appears that the use of different measures and different data sets may bias the empirical results. One way to check for potential biases is to compare marginal distributions at ages 16 and 18 that connect two different cohorts (see Figure 1) since these marginal distributions are observed twice. The comparison gives acceptable results (could be derived from Tables A1 and A2). It should also be noted that in the empirical analysis to be reported later, for all ages except the initial condition (age 13), only transi-

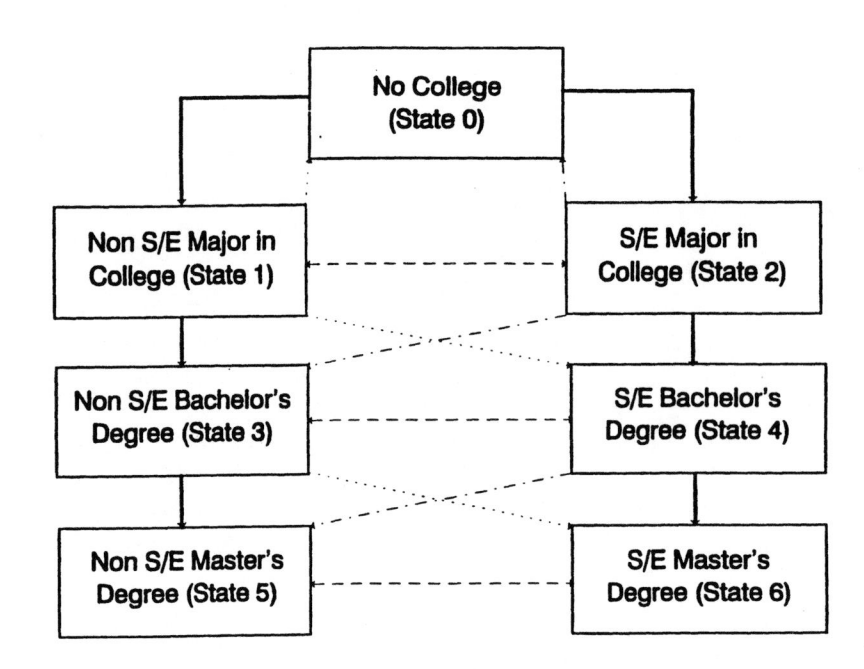

**FIGURE 2** Schematic flows among seven educational states.

*tion* (not marginal) probabilities are used. Marginal distributions of later ages result from the cumulative product of prior transition matrices and the initial condition [as shown in equations (1) and (5)]. Thus, it would be a mistake to attribute the marginal distribution at any given age to the measurement of states at the same age.

## RESULTS

### Distribution of Educational States

Assuming an arbitrary cohort size of 1,000 and the initial state distribution at age 13 to equal that calculated from the LSAY survey, I simulate the distribution of educational states at each age for each sex using equations (1) and (5). Mortality is ignored here. The results of the exercise are presented in Table 1. For example, there are 5.4 and 18.3 females in a 1,000-member cohort whose highest degrees at age 32 are respectively master's and bachelor's degrees in science. The comparable figures are 12.0 and 40.5 for males. These statistics are close to cross-sectional statistics reported elsewhere (National Science Board, 1986).

### Age Pyramid

In my earlier work (Xie, 1989), I hypothesized that the S/E pool forms an age pyramid, especially in early ages: for a given cohort, the pool depletes with age. This is true even though some youth move into the pool while some youth move out of the pool at the same time, because the exit rate exceeds the entry rate.[9]

Such an age pyramid indeed exists. After states (2), (4), and (6) are combined as the S/E pool, the pyramid is shown in Figure 3. The number of cohort members with S/E education steadily decreases from age 13 to age 23, and the decrease is faster for females than for males. After age 23, the pool increases slightly due to a small proportion of cohort members who start or resume postsecondary education after some disruption. This can be seen by the decrease in the number of cohort members in state (0) after age 23.

### Sex Differences

One of the objectives of this study is to know where in the S/E education pipeline females fall behind males. It has been frequently speculated in the literature that females are severely disadvantaged relative to males in

**TABLE 1** Distribution of Educational States by Age and Sex for a Synthetic Cohort

| Age | School Age | 0 | 1 | 2 | 3 | 4 | 5 | 6 | Total |
|---|---|---|---|---|---|---|---|---|---|
| \multicolumn{10}{l}{Panel A: Females} |
| 13 | Grade 7 | 320.0 | 514.0 | 166.0 | 0.0 | 0.0 | 0.0 | 0.0 | 1000 |
| 14 | Grade 8 | 363.3 | 498.6 | 138.2 | 0.0 | 0.0 | 0.0 | 0.0 | 1000 |
| 15 | Grade 9 | 311.8 | 544.6 | 143.7 | 0.0 | 0.0 | 0.0 | 0.0 | 1000 |
| 16 | Grade 10 | 321.2 | 545.6 | 133.2 | 0.0 | 0.0 | 0.0 | 0.0 | 1000 |
| 17 | Grade 11 | 406.0 | 476.5 | 117.5 | 0.0 | 0.0 | 0.0 | 0.0 | 1000 |
| 18 | Grade 12 | 478.0 | 469.3 | 52.7 | 0.0 | 0.0 | 0.0 | 0.0 | 1000 |
| 19 | College 1 | 445.0 | 518.7 | 36.3 | 0.0 | 0.0 | 0.0 | 0.0 | 1000 |
| 20 | College 2 | 541.0 | 429.1 | 29.9 | 0.0 | 0.0 | 0.0 | 0.0 | 1000 |
| 21 | College 3 | 617.6 | 352.3 | 30.2 | 0.0 | 0.0 | 0.0 | 0.0 | 1000 |
| 22 | College 4 | 705.3 | 268.8 | 26.0 | 0.0 | 0.0 | 0.0 | 0.0 | 1000 |
| 23 | Postgraduate 1 | 740.2 | 83.7 | 7.0 | 152.6 | 16.5 | 0.0 | 0.0 | 1000 |
| 26 | Postgraduate 4 | 706.4 | 69.5 | 5.6 | 172.4 | 22.4 | 21.6 | 2.1 | 1000 |
| 32 | Postgraduate 11 | 564.8 | 174.0 | 10.3 | 169.4 | 18.3 | 57.7 | 5.4 | 1000 |
| Panel B: Males | | | | | | | | | |
| 13 | Grade 7 | 380.0 | 425.0 | 195.0 | 0.0 | 0.0 | 0.0 | 0.0 | 1000 |
| 14 | Grade 8 | 422.0 | 382.0 | 196.0 | 0.0 | 0.0 | 0.0 | 0.0 | 1000 |
| 15 | Grade 9 | 365.8 | 478.1 | 156.1 | 0.0 | 0.0 | 0.0 | 0.0 | 1000 |
| 16 | Grade 10 | 420.2 | 454.5 | 125.3 | 0.0 | 0.0 | 0.0 | 0.0 | 1000 |
| 17 | Grade 11 | 480.5 | 394.5 | 125.0 | 0.0 | 0.0 | 0.0 | 0.0 | 1000 |
| 18 | Grade 12 | 520.4 | 380.9 | 98.7 | 0.0 | 0.0 | 0.0 | 0.0 | 1000 |
| 19 | College 1 | 479.2 | 418.2 | 102.6 | 0.0 | 0.0 | 0.0 | 0.0 | 1000 |
| 20 | College 2 | 532.5 | 375.4 | 92.1 | 0.0 | 0.0 | 0.0 | 0.0 | 1000 |
| 21 | College 3 | 582.9 | 320.4 | 96.7 | 0.0 | 0.0 | 0.0 | 0.0 | 1000 |
| 22 | College 4 | 681.9 | 248.4 | 69.7 | 0.0 | 0.0 | 0.0 | 0.0 | 1000 |
| 23 | Postgraduate 1 | 739.3 | 88.3 | 19.5 | 111.5 | 41.3 | 0.0 | 0.0 | 1000 |
| 26 | Postgraduate 4 | 692.5 | 68.1 | 20.8 | 140.4 | 53.5 | 17.9 | 6.9 | 1000 |
| 32 | Postgraduate 11 | 569.3 | 136.0 | 35.6 | 141.1 | 40.5 | 65.5 | 12.0 | 1000 |

NOTE: Educational states are defined as: (0) secondary education or college dropout, (1) intended or actual non-S/E postsecondary education, (2) intended or actual S/E postsecondary education, (3) non-S/E bachelor's degree, (4) S/E bachelor's degree, (5) non-S/E master's degree, and (6) S/E master's degree.

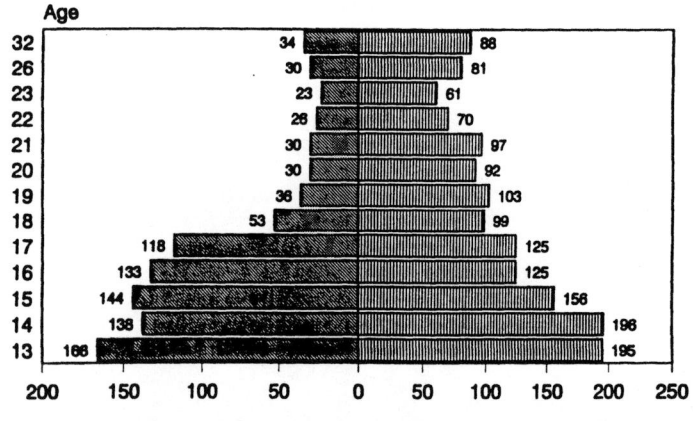

CAREERS IN SCIENCE AND TECHNOLOGY: AN INTERNATIONAL PERSPECTIVE

**FIGURE 3** Age pyramid of cohort members with S/E education in a 1000-member cohort.

SOURCE: Sums of columns 2, 4, and 6 in Table 1.

early ages. In the executive summary of her report to the Rockefeller Foundation, for example, Berryman (1983) remarked that "strategies to increase the size of the initial scientific/mathematical pool of minorities and women should be targeted before and during high school" (p.7).

Sex ratio, a simple measure of the composition by sex, is defined as the number of females to the number of males. Figure 4 depicts the sex ratio measure in the S/E pool by S/E educational state and age. It is evident that the greatest drops in sex ratio occur between ages 17 and 18, around high school graduation. It should be pointed out that this new finding cannot be simply attributed to the different measures used for the LSAY1 and NLS-72 cohorts, for the *same* measures are used for both males and females within each cohort.[10] In addition, the most drastic drop for females occurs between ages 17 and 18 (117.5 to 52.7 per 1,000, Table 1), *within* the same LSAY1 cohort. One explanation is that young women of ages 17-18 are more knowledgeable about, or more influenced by, than ever before, occupational sex segregation in the labor force, while having to make realistic decisions about career, marriage, and family. Thus, around high school graduation, many more women than in earlier years are discouraged from pursuing S/E careers.

Equation (5) means that the process of becoming a scientist/engineer is a cumulative process. What happens at each stage contributes to the final outcome. Thus, a large change at one single stage may only have a small ripple effect on the final outcome. In Table 2,

I sequentially substitute males' transition rates for females' at different ages to reveal the effects of the substitution on narrowing the gender gap in educational outcome at age 32. Surprisingly, the effects are all small prior to college years.

Next, let us ask the following counter-factual question: How much reduction in sex differences at age 32 would occur if there were no sex differences in initial distribution of educational states at an earlier age? To answer this question, I forced gender equality by assigning males' marginal distribution of educational states to females at ages prior to 32 and observe changes in the gender gap at age 32. The results of this exercise are reported in Table 3. Each line in Table 3 represents the remaining gender gap in educational states at age 32 after the marginal distribution is equalized at an earlier age. For example, sex differences in high school contribute little (between 0.1 and 2.2 percent) to the sex differences in outcomes at age 32. However, these results, combined with those from Table 2, do not mean that sex differences before age 19 were small. As we have seen in Figure 4, the differences are large, especially at age 18. The results simply mean that the large sex differences observed in high school years are absorbed by sex differences in the educational process of college and graduate years. That is, if conditions in high school were radically changed so that females and males were equally interested in pursuing S/E careers, women would still be underrepresented as long as conditions remain the same in later years.

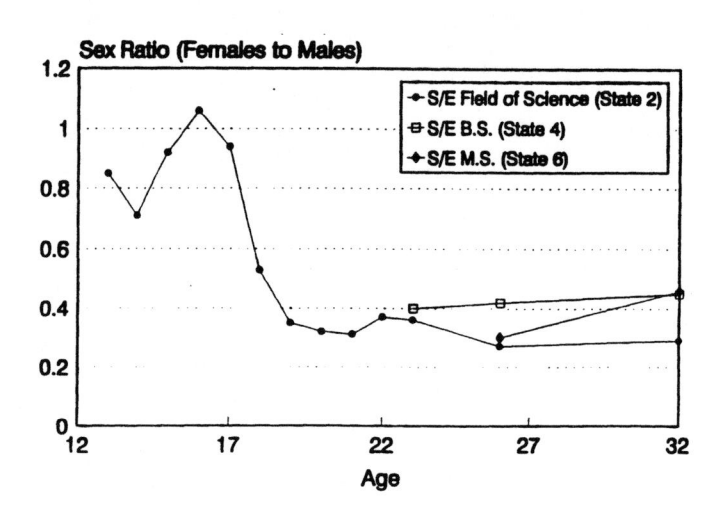

SOURCE: Columns 2, 4, and 6 in Table 1.

**FIGURE 4** Sex ratio of cohort members with S/E education by age and educational states.

**TABLE 2** Number of Females in Science/Engineering Educational States at Age 32 if Male's Transition Rates Were True for Females at Different Ages

| Males' Rates were Used at Age | School Age | 2 | % Explained Gender Gap | 4 | % Explained Gender Gap | 6 | % Explained Gender Gap |
|---|---|---|---|---|---|---|---|
| No Substitution (Observed) | | 10.3 | | 18.3 | | 5.4 | |
| 13 | Grade 7 | 10.3 | 0.0% | 18.3 | 0.0% | 5.4 | 0.0% |
| 14 | Grade 8 | 10.3 | 0.0 | 18.3 | -0.2 | 5.4 | -0.2 |
| 15 | Grade 9 | 10.3 | 0.1 | 18.2 | -0.8 | 5.4 | -0.9 |
| 16 | Grade 10 | 10.3 | 0.0 | 18.4 | 0.1 | 5.4 | -0.1 |
| 17 | Grade 11 | 10.3 | 0.0 | 19.0 | 2.9 | 5.6 | 1.7 |
| 18 | Grade 12 | 10.3 | -0.1 | 20.2 | 8.4 | 5.8 | 4.9 |
| 19 | College 1 | 10.2 | -0.3 | 19.9 | 6.9 | 5.8 | 5.1 |
| 20 | College 2 | 10.2 | -0.5 | 22.9 | 20.7 | 6.3 | 13.4 |
| 21 | College 3 | 10.3 | -0.1 | 20.9 | 11.8 | 5.9 | 7.0 |
| 22 | College 4 | 10.7 | 1.7 | 21.1 | 12.6 | 5.3 | -1.7 |
| 23 | Postgraduate 1 | 11.8 | 5.8 | 24.1 | 25.8 | 6.8 | 21.5 |
| 26 | Postgraduate 4 | 33.4 | 91.0 | 20.6 | 10.4 | 6.8 | 21.0 |

NOTE: Educational states are defined as: (0) secondary education or college dropout, (1) intended or actual non-S/E postsecondary education, (2) intended or actual S/E postsecondary education, (3) non-S/E bachelor's degree, (4) S/E bachelor's degree, (5) non-S/E master's degree, and (6) S/E master's degree. The exercise alternately substitutes one set of males' transition rates at a given age while keeping everything else intact.

## Exit and Entry Rates

Given our simple "bathtub" model, women's underrepresentation in S/E educational states could result from two sources: women's exit rate from S/E states is higher than men's or women's entry rate into S/E states is lower than men's. Two series of exit and entry rates are displayed in Figures 5 and 6.[11] In calculating the rates, I combined the three S/E states into a single state and lumped the four non-S/E states into another state. An interesting result is that the exit rate for females trails that for males closely except at age 17. A larger and more consistent gender gap, however, is observed for the entry rate after age 17. From these two figures, I infer that a large portion of the gender gap in attaining S/E education is not merely due to women's higher likelihood to exit the S/E pool. Men are just as likely as women to drop out of the S/E pool, but their likelihood to enter or re-enter the pool once out of it is significantly higher, particularly in later years. Unfortunately, past research has not paid attention to this problem. For example, my results contradict Berryman's (1983, p.7) assertion that "*after high school, migration is almost entirely out of, not into, the pool*" (emphasis original). For males, I have found that the rate of migration into the pool is around 4 percent, compared to less than 2 percent for women.

As small as they may seem, cumulatively these figures are very significant given that only 2 percent of females and 5 percent of males obtain bachelor's and master's degrees in science by age 32 (see Table 1).

## S/E Occupations

Education affects occupation, but only in non-deterministic ways. Obtaining S/E education means that one's likelihood of working in a S/E occupation significantly increases, but it cannot be equated with S/E occupation. In fact, Table 4 shows that only 9 percent of females and 35 percent of males with S/E bachelor's degrees have S/E occupations at age 32. At the master's level, the percentages are better, 27 percent for females and 54 percent for males.

In Table 5, I present simulated occupational distributions at age 32 under four different conditions.[12]

**TABLE 3** Number of Females in S/E Educational States at Age 32 if Male's Distribution Were True for Females at Different Ages

| | | Educational State | | | | | |
|---|---|---|---|---|---|---|---|
| Males' Distribution were Used at Age | School Age | 2 | % Explained Gender Gap | 4 | % Explained Gender Gap | 6 | % Explained Gender Gap |
| No Substitution (Observed) | | 10.3 | 18.3 | 5.4 | | | |
| 13 | Grade 7 | 10.3 | 0.0% | 18.3 | 0.0% | 5.4 | 0.0% |
| 14 | Grade 8 | 10.3 | 0.0 | 18.3 | -0.1 | 5.4 | -0.1 |
| 15 | Grade 9 | 10.3 | 0.0 | 18.3 | -0.2 | 5.4 | -0.3 |
| 16 | Grade 10 | 10.3 | 0.1 | 18.1 | -1.1 | 5.4 | -1.2 |
| 17 | Grade 11 | 10.3 | 0.2 | 18.1 | -0.9 | 5.4 | -1.2 |
| 18 | Grade 12 | 10.3 | 0.1 | 18.8 | 2.2 | 5.5 | 0.8 |
| 19 | College 1 | 10.3 | 0.0 | 20.6 | 10.1 | 5.8 | 5.4 |
| 20 | College 2 | 10.2 | -0.3 | 22.3 | 18.0 | 6.2 | 11.0 |
| 21 | College 3 | 10.1 | -0.7 | 26.3 | 35.9 | 6.9 | 22.5 |
| 22 | College 4 | 10.1 | -0.7 | 28.4 | 45.4 | 7.2 | 27.5 |
| 23 | Postgraduate 1 | 10.6 | 1.1 | 30.7 | 55.5 | 7.0 | 24.2 |
| 26 | Postgraduate 4 | 12.0 | 6.9 | 37.0 | 84.1 | 9.0 | 54.3 |

NOTE: Educational states are defined as: (0) secondary education or college dropout, (1) intended or actual non-S/E postsecondary education, (2) intended or actual S/E postsecondary education, (3) non-S/E bachelor's degree, (4) S/E bachelor's degree, (5) non-S/E master's degree, and (6) S/E master's degree. The exercise alternately substitutes males' educational distribution at a given age while keeping everything else intact.

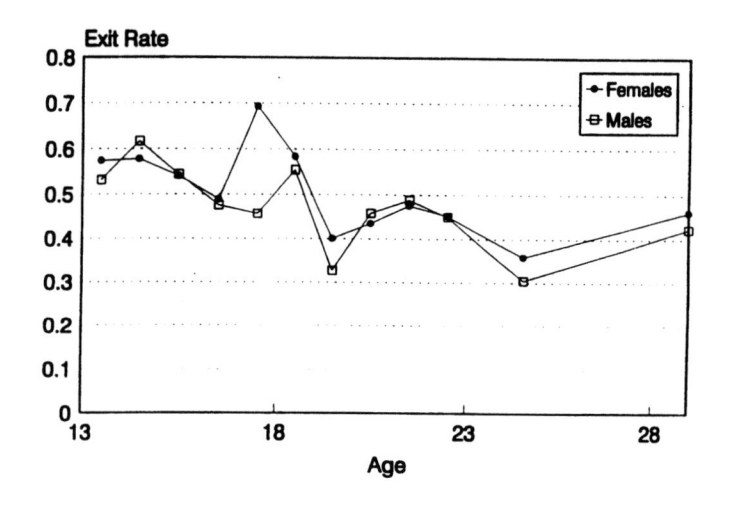

SOURCE: LSAY and NLS-72.

**FIGURE 5** Exit rate from S/E educational pool by age and sex.

SOURCE: LSAY and NLS-72.

**FIGURE 6** Entry rate into S/E educational pool by age and sex.

In Panel A, females' educational distribution at age 32 (last line, Panel A of Table 4) is used. In Panel B, males' educational distribution at age 32 (last line, Panel B of Table 4) is used. Within each panel of Table 5, two lines represent two sets of transition rates from educational states to occupational states, one for females and one for males. Thus, the first line of Panel A and the second line of Panel B are simulated distributions respectively for females and males using gender-specific information. That is, for a 1,000-member female cohort following the age-specific transition rates observed for our data, only 14.5 of them work as scientists/engineers at age 32. The comparable number is 59.5 for a 1,000-member male cohort. Most of the gender gap is due to women's lower likelihood to work as scientists/engineers given the same educational background. If females had the same distribution of educational states as males, female scientists/engineers would increase to 17.5 per 1,000 (first line of Panel B). However, if females had the same Table 4 transition rates from educational states to occupational states at rates of transition from education to occupation as males, female scientists/engineers would increase to 49.8 per 1,000. Therefore, women's lower achievement in attaining S/E education can only explain a small fraction (about 10 percent) of the gender gap in attaining S/E occupations.

## CONCLUSION

In brief, I offer the following conclusions:

1. This paper proposes a new demographic approach to studying the process of becoming a scientist/engineer.

2. The proposed approach consists of constructing a synthetic cohort from different longitudinal surveys and modeling the career process as a Markov process as the cohort ages (or matures).

3. An age pyramid in the S/E educational pool is found to exist. Generally speaking, for any given cohort, the proportion of people remaining in the pool decreases with age.

4. Women's representation in the S/E educational pool drops suddenly near high school graduation (between ages 17 and 18).

5. College and graduate years account for most of the sex differences in the proportion attaining science degrees by early adulthood.

6. Men and women differ more in the entry or re-entry rate into the S/E educational pool than the exit rate out of the S/E educational pool.

7. Women's underrepresentation in S/E occupations is mainly due to women's lower likelihood of being employed in S/E occupations net of differential access to S/E education rather than women's lower likelihood of having a S/E education.

**TABLE 4** Transition Rates From Educational States to Occupational States at Age 32 by Sex

| | Occupational States | | | |
|---|---|---|---|---|
| Educational State | Not Working | Non S/E | S/E | (n) |
| **Panel A: Females** | | | | |
| Secondary education only | 34.41% | 65.09 | 0.50 | (3583) |
| Non-S/E postsecondary education | 25.32 | 72.61 | 2.07 | (1110) |
| S/E postsecondary education | 16.67 | 81.67 | 1.67 | (60) |
| Non-S/E bachelor's degree | 25.92 | 71.84 | 2.24 | (1115) |
| S/E bachelor's degree | 35.34 | 55.64 | 9.02 | (133) |
| Non-S/E master's degree | 16.16 | 82.17 | 1.67 | (359) |
| S/E master's degree | 36.36 | 36.36 | 27.27 | (33) |
| **Panel B: Males** | | | | |
| Secondary education only | 12.64% | 84.97 | 2.39 | (3354) |
| Non-S/E postsecondary education | 15.14 | 79.63 | 5.22 | (766) |
| S/E postsecondary education | 15.67 | 77.42 | 6.91 | (217) |
| Non-S/E bachelor's degree | 4.57 | 87.45 | 7.98 | (940) |
| S/E bachelor's degree | 6.38 | 57.38 | 36.24 | (298) |
| Non-S/E master's degree | 7.13 | 86.94 | 5.94 | (421) |
| S/E master's degree | 5.06 | 40.51 | 54.43 | (79) |

NOTE: Main entries are row percentages, i.e., estimated probability of occupational states conditional on educational states.

**TABLE 5** Simulated Occupational Distribution at Age 32 Under Different Assumptions

| Educational State | Occupational States | | | |
|---|---|---|---|---|
| | Not Working | S/E | Non S/E | (n) |
| **Panel A: Females' Educational Distribution is Used (Last Line, Panel A of Table 1)** | | | | |
| Females' Transition Rates are Used (Panel A of Table 4) | 301.8 | 683.6 | 14.5 | (1000) |
| Males' Transition Rates are Used (Panel B of Table 4) | 112.7 | 837.5 | 49.8 | (1000) |
| **Panel B: Males' Educational Distribution is Used (Last Line, Panel B of Table 1)** | | | | |
| Females' Transition Rates are Used (Panel A of Table 4) | 302.1 | 680.4 | 17.5 | (1000) |
| Males' Transition Rates are Used (Panel B of Table 4) | 112.5 | 828.0 | 59.5 | (1000) |

NOTE: Rates of transition from education to occupation by sex are reported in Table 4.

## NOTES

1. As Elder (1985) recognizes, the concept of career has been generalized to the concept of trajectory in the literature on life course.

2. This is akin to Quetelet's (1842) notion of the "average man," or the "social man." One major difference, however, is that I aim merely at characterizing a population or a subpopulation, whereas Quetelet's objective was to capture the "essence" of a society.

3. One major difference between what I propose here and conventional multistate life tables is that age is used here as a truly discrete variable (actually school age) whereas it is usually used as a continuous variable in multistate life tables. Using age as a discrete variable is legitimate in the present case given that transitions in school take discrete jumps annually (grade or class).

4. The 1986 follow-up was a nonrandom subsample with full coverage of college graduates. Other sample members who were retained with certainty included Hispanics, teachers and potential teachers, and persons who were divorced, widowed or separated from their spouses, or never-married parents. This problem is handled by weighing observations by the inverse of the probability of being included.

5. Biological age is actually an approximation, translated from school age as shown in Table 1.

6. Social science fields are excluded from this study. The detailed codes and titles are available upon request.

7. There are other possible flows (such as jumping from not being in college to having a master's degree) that are observed in Tables A1 and A2 due to lack of data between two observation periods.

8. Since the sample sizes are large for all transition matrices, sampling errors are ignored in the analysis.

9. See also Berryman, 1983 (p. 4).

10. Unless, of course, it can be shown that at least one of the measures is gender-biased.

11. Midpoints of transition intervals are used for the horizontal axis (age) in Figures 5 and 6.

12. S/E occupations were recorded from detailed three-digit 1970 census occupational codes.

I thank Paul Baltes, Chris Bettinger, William Frey, Kenneth Land, Scott Long, Rob Mare, Jon Miller, Andrei Rogers, and Peter Tiemeyer for their comments and/or advice.

**TABLE A1** Matrices of Transition Among Educational States, Females

| School Ages | Origin State | 0 | 1 | 2 | 3 | 4 | 5 | 6 | (n) |
|---|---|---|---|---|---|---|---|---|---|
| Grade 7 | 0 | 0.6219 | 0.3060 | 0.0721 | -- | -- | -- | -- | (402) |
| to Grade 8 | 1 | 0.2407 | 0.6728 | 0.0864 | -- | -- | -- | -- | (648) |
|  | 2 | 0.2440 | 0.3301 | 0.4258 | -- | -- | -- | -- | (209) |
| Grade 8 | 0 | 0.5718 | 0.3536 | 0.0746 | -- | -- | -- | -- | (362) |
| to Grade 9 | 1 | 0.1676 | 0.7156 | 0.1168 | -- | -- | -- | -- | (531) |
|  | 2 | 0.1481 | 0.4296 | 0.4222 | -- | -- | -- | -- | (135) |
| Grade 9 | 0 | 0.6532 | 0.2626 | 0.0842 | -- | -- | -- | -- | (297) |
| to Grade 10 | 1 | 0.1598 | 0.7648 | 0.0754 | -- | -- | -- | -- | (557) |
|  | 2 | 0.2123 | 0.3288 | 0.4589 | -- | -- | -- | -- | (146) |
| Grade 10 | 0 | 0.8220 | 0.1602 | 0.0178 | -- | -- | -- | -- | (337) |
| to Grade 11 | 1 | 0.2313 | 0.6886 | 0.0801 | -- | -- | -- | -- | (562) |
|  | 2 | 0.1185 | 0.3704 | 0.5111 | -- | -- | -- | -- | (135) |
| Grade 11 | 0 | 0.8410 | 0.1503 | 0.0087 | -- | -- | -- | -- | (346) |
| to Grade 12 | 1 | 0.2450 | 0.7275 | 0.0275 | -- | -- | -- | -- | (400) |
|  | 2 | 0.1683 | 0.5248 | 0.3069 | -- | -- | -- | -- | (101) |
| Grade 12 | 0 | 0.7767 | 0.2140 | 0.0093 | -- | -- | -- | -- | (3883) |
| to College 1 | 1 | 0.1455 | 0.8332 | 0.0213 | -- | -- | -- | -- | (3196) |
|  | 2 | 0.1034 | 0.4803 | 0.4163 | -- | -- | -- | -- | (406) |
| College 1 | 0 | 0.9000 | 0.0963 | 0.0037 | -- | -- | -- | -- | (5069) |
| to College 2 | 1 | 0.2576 | 0.7300 | 0.0124 | -- | -- | -- | -- | (5012) |
|  | 2 | 0.1910 | 0.2095 | 0.5995 | -- | -- | -- | -- | (377) |
| College 2 | 0 | 0.8966 | 0.0976 | 0.0058 | -- | -- | -- | -- | (5715) |
| to College 3 | 1 | 0.2973 | 0.6790 | 0.0237 | -- | -- | -- | -- | (4016) |
|  | 2 | 0.1644 | 0.2705 | 0.5651 | -- | -- | -- | -- | (292) |
| College 3 | 0 | 0.9285 | 0.0666 | 0.0048 | -- | -- | -- | -- | (5988) |
| to College 4 | 1 | 0.3506 | 0.6291 | 0.0203 | -- | -- | -- | -- | (2313) |
|  | 2 | 0.2757 | 0.2000 | 0.5243 | -- | -- | -- | -- | (185) |
| College 4 | 0 | 0.9036 | 0.0757 | 0.0079 | 0.0119 | 0.0009 | -- | -- | (6565) |
| to Postgraduate 1 | 1 | 0.3517 | 0.1052 | 0.0036 | 0.5322 | 0.0072 | -- | -- | (1939) |
|  | 2 | 0.3239 | 0.0795 | 0.0170 | 0.0455 | 0.5341 | -- | -- | (176) |
| Postgraduate 1 | 0 | 0.8992 | 0.0766 | 0.0057 | 0.0161 | 0.0015 | 0.0008 | -- | (6093) |
| to Postgraduate 4 | 1 | 0.4523 | 0.1454 | 0.0088 | 0.3604 | 0.0183 | 0.0142 | 0.0007 | (1479) |
|  | 2 | 0.4179 | 0.0821 | 0.0896 | 0.0597 | 0.3358 | 0.0149 | 0.0000 | (134) |
|  | 3 | -- | -- | -- | 0.8387 | 0.0335 | 0.1254 | 0.0024 | (1252) |
|  | 4 | -- | -- | -- | 0.1159 | 0.7464 | 0.0362 | 0.1014 | (138) |
| Postgraduate 4 | 0 | 0.7712 | 0.1992 | 0.0097 | 0.0169 | 0.0013 | 0.0018 | 0.0000 | (3916) |
| to Postgraduate 11 | 1 | 0.2689 | 0.4499 | 0.0391 | 0.2029 | 0.0122 | 0.0220 | 0.0049 | (409) |
|  | 2 | 0.2308 | 0.3590 | 0.1282 | 0.1795 | 0.0513 | 0.0256 | 0.0256 | (39) |
|  | 3 | -- | -- | -- | 0.7990 | 0.0174 | 0.1732 | 0.0104 | (1149) |
|  | 4 | -- | -- | -- | 0.2098 | 0.5944 | 0.1608 | 0.0350 | (143) |
|  | 5 | -- | -- | -- | -- | -- | 0.9426 | 0.0574 | (122) |
|  | 6 | -- | -- | -- | -- | -- | 0.4545 | 0.5455 | (11) |

NOTE: Main entries are row proportions. Educational states are defined as: (0) secondary education only, (1) intended or actual non-S/E postsecondary education, (2) intended or actual S/E postsecondary education, (3) non-S/E bachelor's degree, (4) S/E bachelor's degree, (5) non-S/E master's degree, and (6) S/E master's degree. Cells omitted and marked with "--" are structural zeros.

**TABLE A2** Matrices of Transition Among Educational States, Males

| School Ages | Origin State | 0 | 1 | 2 | 3 | 4 | 5 | 6 | (n) |
|---|---|---|---|---|---|---|---|---|---|
| Grade 7 | 0 | 0.6730 | 0.2409 | 0.0860 | -- | -- | -- | -- | (523) |
| to Grade 8 | 1 | 0.2940 | 0.5368 | 0.1692 | -- | -- | -- | -- | (585) |
|  | 2 | 0.2119 | 0.3197 | 0.4684 | -- | -- | -- | -- | (269) |
| Grade 8 | 0 | 0.6284 | 0.2775 | 0.0940 | -- | -- | -- | -- | (436) |
| to Grade 9 | 1 | 0.2047 | 0.6871 | 0.1082 | -- | -- | -- | -- | (425) |
|  | 2 | 0.1144 | 0.5025 | 0.3831 | -- | -- | -- | -- | (201) |
| Grade 9 | 0 | 0.7859 | 0.1746 | 0.0394 | -- | -- | -- | -- | (355) |
| to Grade 10 | 1 | 0.2370 | 0.6798 | 0.0832 | -- | -- | -- | -- | (481) |
|  | 2 | 0.1243 | 0.4201 | 0.4556 | -- | -- | -- | -- | (169) |
| Grade 10 | 0 | 0.8174 | 0.1545 | 0.0281 | -- | -- | -- | -- | (356) |
| to Grade 11 | 1 | 0.2511 | 0.6444 | 0.1044 | -- | -- | -- | -- | (450) |
|  | 2 | 0.1829 | 0.2927 | 0.5244 | -- | -- | -- | -- | (164) |
| Grade 11 | 0 | 0.8113 | 0.1606 | 0.0282 | -- | -- | -- | -- | (355) |
| to Grade 12 | 1 | 0.2886 | 0.6676 | 0.0437 | -- | -- | -- | -- | (343) |
|  | 2 | 0.1339 | 0.3228 | 0.5433 | -- | -- | -- | -- | (127) |
| Grade 12 | 0 | 0.7890 | 0.1629 | 0.0482 | -- | -- | -- | -- | (3696) |
| to College 1 | 1 | 0.1545 | 0.7573 | 0.0882 | -- | -- | -- | -- | (2505) |
|  | 2 | 0.0984 | 0.4561 | 0.4455 | -- | -- | -- | -- | (1138) |
| College 1 | 0 | 0.8762 | 0.1008 | 0.0230 | -- | -- | -- | -- | (4960) |
| to College 2 | 1 | 0.2251 | 0.7460 | 0.0289 | -- | -- | -- | -- | (4225) |
|  | 2 | 0.1806 | 0.1466 | 0.6727 | -- | -- | -- | -- | (1207) |
| College 2 | 0 | 0.8640 | 0.1058 | 0.0303 | -- | -- | -- | -- | (5352) |
| to College 3 | 1 | 0.2712 | 0.6470 | 0.0818 | -- | -- | -- | -- | (3669) |
|  | 2 | 0.2278 | 0.2307 | 0.5415 | -- | -- | -- | -- | (1023) |
| College 3 | 0 | 0.9203 | 0.0670 | 0.0127 | -- | -- | -- | -- | (5360) |
| to College 4 | 1 | 0.3449 | 0.6148 | 0.0403 | -- | -- | -- | -- | (1960) |
|  | 2 | 0.3611 | 0.1281 | 0.5108 | -- | -- | -- | -- | (601) |
| College 4 | 0 | 0.9000 | 0.0763 | 0.0181 | 0.0035 | 0.0022 | -- | -- | (5978) |
| to Postgraduate 1 | 1 | 0.4058 | 0.1272 | 0.0160 | 0.4320 | 0.0190 | -- | -- | (1683) |
|  | 2 | 0.3567 | 0.0678 | 0.0460 | 0.0263 | 0.5033 | -- | -- | (457) |
| Postgraduate 1 | 0 | 0.8823 | 0.0738 | 0.0226 | 0.0134 | 0.0071 | 0.0007 | 0.0002 | (5665) |
| to Postgraduate 4 | 1 | 0.3883 | 0.1376 | 0.0227 | 0.3860 | 0.0408 | 0.0222 | 0.0023 | (1715) |
|  | 2 | 0.3020 | 0.0720 | 0.1060 | 0.0700 | 0.4300 | 0.0060 | 0.0140 | (500) |
|  | 3 | -- | -- | -- | 0.8288 | 0.0450 | 0.1205 | 0.0056 | (888) |
|  | 4 | -- | -- | -- | 0.0617 | 0.7560 | 0.0456 | 0.1367 | (373) |
| Postgraduate 4 | 0 | 0.7831 | 0.1543 | 0.0415 | 0.0131 | 0.0046 | 0.0035 | 0.0000 | (3448) |
| to Postgraduate 11 | 1 | 0.3405 | 0.3432 | 0.0429 | 0.2038 | 0.0107 | 0.0456 | 0.0134 | (373) |
|  | 2 | 0.1844 | 0.2766 | 0.1915 | 0.0851 | 0.2411 | 0.0071 | 0.0142 | (141) |
|  | 3 | -- | -- | -- | 0.7740 | 0.0181 | 0.1951 | 0.0128 | (938) |
|  | 4 | -- | -- | -- | 0.1459 | 0.5431 | 0.2488 | 0.0622 | (418) |
|  | 5 | -- | -- | -- | -- | -- | 0.9694 | 0.0306 | (98) |
|  | 6 | -- | -- | -- | -- | -- | 0.2619 | 0.7381 | (42) |

NOTE: Main entries are row proportions. Educational states are defined as: (0) secondary education only, (1) intended or actual non-S/E postsecondary education, (2) intended or actual S/E postsecondary education, (3) non-S/E bachelor's degree, (4) S/E bachelor's degree, (5) non-S/E master's degree, and (6) S/E master's degree. Cells omitted and marked with "--" are structural zeros.

## REFERENCES

Bartholomew, D. J. 1973. Stochastic Models for Social Processes, 2nd edition. London: Wiley.

Berryman, S. E. 1983. Who Will Do Science? A special Report, the Rockefeller Foundation.

Brown, D., L. Brooks and Associates (eds). 1990. Career Choice and Development, 2nd edition. San Francisco: Jossey-Bass Publishers.

Elder, G. H., Jr. 1985. Perspectives on the Life Course. Pages 23-49 in Life Course Dynamics: Trajectories and Transitions, 1968-1980, edited by Glen H. Elder, Jr. Ithaca, NY: Cornell University Press.

Freeman, R. B. 1971. The Market for College-Trained Manpower: A Study in the Economics of Career Choice. Cambridge, MA: Harvard University Press.

Ginzberg, E., S. W. Ginsburg, S. Axelrad, and J.L. Herma. 1951. Occupational Choice: An Approach to a General Theory. New York: Columbia University Press.

Huff, R. A. and M. O. Chandler. 1970. A Taxonomy of Instructional Programs in Higher Education. Washington, D.C.: U.S. Printing Office.

Hoem, J. M. 1977. A Markov Chain Model of Working Life Tables. Scandinavian Actuarial Journal, 1-20.

Land, K. C. and G. C. Hough, Jr. 1989. New Methods for Tables of School Life, With Applications to U.S. Data from Recent School Years. Journal of the American Statistical Association, 84:63-75.

Malitz, G. S. 1981. A Classification of Instructional Programs. Washington, D.C.: U.S. Printing Office.

Namboodiri, K. and C. M. Suchindran. 1987. Life Table Techniques and their Applications. Orlando, FL: Academic Press.

National Science Board. 1986. Undergraduate Science, Mathematics and Engineering Education. Report of the NSB Task Committee on Undergraduate Science and Engineering Education.

Quetelet, L. A. J. 1842. A Treatise on Man and the Development of his Faculties, a facsimile reproduction of the English translation of 1942. Gainesville, FL: Scholar's Facsimiles and Reprints, 1969.

Ryder, N. B. 1965. The Cohort as a Concept in the Study of Social Change. American Sociological Review, 30:843-861.

Sewell, W. H., A. O. Haller, and A. Portes. 1969. The Educational and Early Occupational Attainment Process. American Sociological Review, 34:82-92.

Tuma, N. B. and M. T. Hannan. 1984. Social Dynamics: Models and Methods. San Francisco: Academic Press.

Xie, Y. 1989. The Process of Becoming a Scientist. Unpublished Dissertation, University of Wisconsin.

# Human Resources in Science and Technology and the Less Developed Countries Of Europe (EC-12)

## Emilio Muñoz[1]

The study of human capital is a crucial issue in the attempt to develop regional strategies that increase the potential capabilities of the less favored regions in regard to their own social and economic progression. However, the knowledge of the problem has been hampered by its relevance, its intrinsic complexity, and its multifaceted nature.

One reason for this is that the subject does not depend on a single area of policymaking but rather deals with several: education, employment, and economic policy, as well as the influence of scientific and technological policy on economic and social development.

A second reason for the difficulty in treating this issue lies in the diversity, as well as in the complexity, of the indicators needed to afford an exact diagnosis of the problem for further action. In some cases the indicators are too broad to tune up appropriate reflections. In other cases they appear too specialized and/or excessively reliant on old-fashioned taxonomies to gain the right perception.

All of these shortcomings have a particular significance when one attempts to understand the realities hidden under the realm of science and technology in relation to the less developed regions of Europe. We cannot forget that these regions possess characteristics of industrialized countries and, as such, are part of global projects [i.e., for operational purposes like EC-12 or for analytical targets like the Organization for Economic Cooperation and Development (OECD)].

In this study we will cover some of the key topics related to educational and employment policies, albeit in a general way, and later delve into the science and technology issues concerning human resources and their implications on socioeconomic development.

## EDUCATIONAL POLICY

The European countries share, to a great extent, an educational tradition modeled by the deep processes of great historical-cultural events, such as romanticism, Christianization, and the incorporation of realism.

The present educational guide is the result of a combination of four main approaches:

1. The inheritor of medieval tradition that promotes the transmission of the classical idea of culture

2. The approach that sees the encyclopedic view and defends the diffusion of all available knowledge

3. The polytechnic approach that emphasizes the transfer of skills useful for the economy and productivity

4. The pragmatic approach that supports the idea of the preparation of the individual to cope with the problems of ordinary life

This mixed view has taken education to a point

where, in spite of the differences between countries and government, European schools are studying almost the same content in every country.

Broadly speaking, the European curriculum has been built within a framework of knowledge based on the disciplines, where progress is continuously being incorporated, without taking into account the global approach of looking inside the psycho-pedagogy focal point.

Along these lines, the idea of an everlasting educational crisis is present. This is due to the nature of the educational process itself, in which it is necessary to find adequate interplay between the information transmitter (the teacher) and the receiver (the pupil), together with the essential universality of the contents being taught.

It is worth noting, paradoxically, that in a moving field like education, the application of conservative policies appear to give better results than that of innovative policies. This implies that the evolution trend in educational affairs follows a vicious circle.

In spite of the strong momentum given to education in the late 1940s in western Europe, it is generally believed, at the present time, that European education does not fulfill the requirements raised by economic, cultural, technical, and societal needs. This controversial issue has opened up a debate between teaching staff and public administrators about the resources deemed necessary to reach the appropriate targets.

In the developed countries, education is not a first priority as it has to contend with other highly relevant issues like unemployment, the incorporation of young people into the labor force, social protection, and environmental preservation. Consequently, the public expenditure rate on education has been decreasing in most countries (see Figure 1). However, be reminded that the data, as previously mentioned, is too general and, by the lack of disaggregation level, does not allow conclusions with a good degree of acuteness for analyzing regional disequilibria.

In any case, the data shown reveal that public expenditure is smaller in countries with less development and a greater heterogeneity, and present a clear-cut geographical distribution between a highly developed north and a less developed south. Therefore, the expenditure per capita, as illustrated in Table 1, confirms this trend: the positions at the bottom of the table are held by countries with less developed regions. Against this background the change in educational pat-

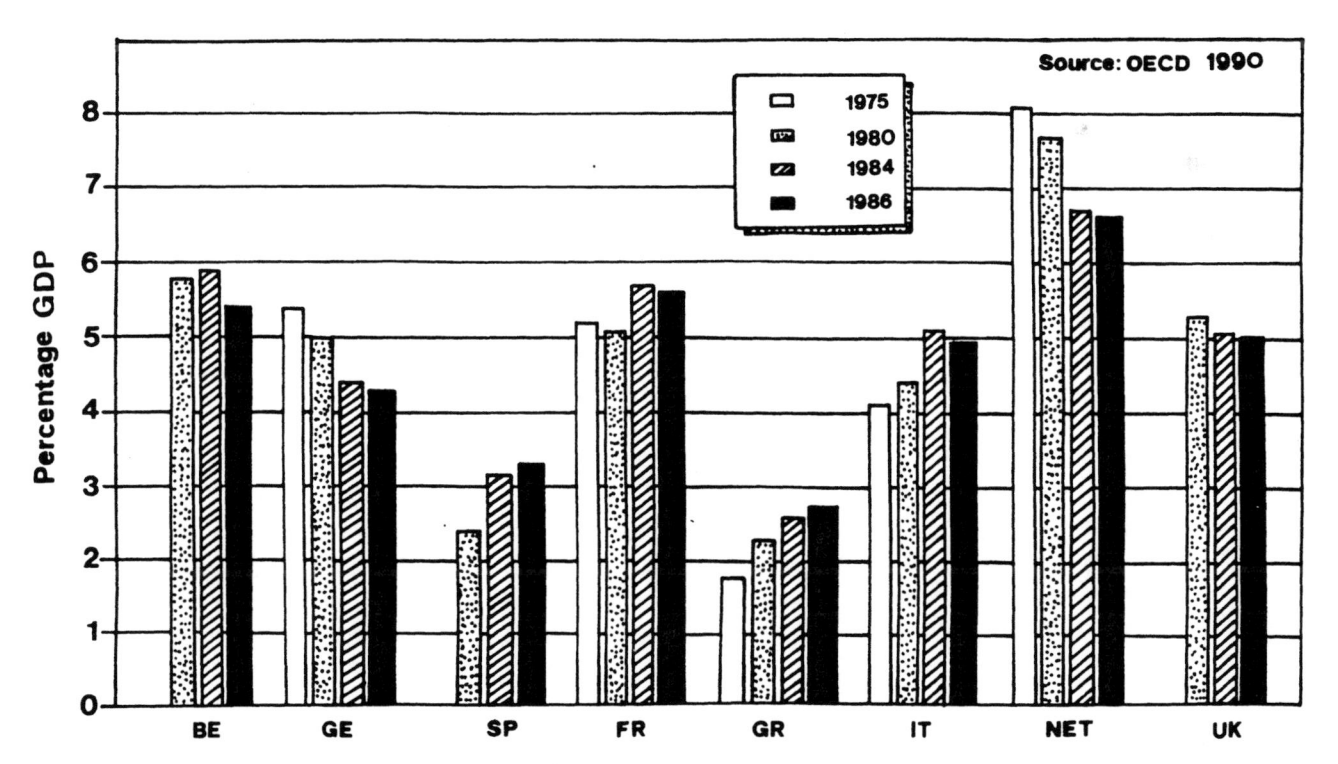

SOURCE: Investigacion y Ciencia, December 1992.

**FIGURE 1** Evolution of public expenditure on education in some community cultures.

**TABLE 1** Public Expenditure on Education per Capita In U.S. Dollars (Purchasing Power Parity) in the EC Countries

| | |
|---|---|
| Denmark | 934 |
| Netherlands | 838 |
| France | 746 |
| Belgium | 745 |
| United Kingdom | 703 |
| Italy | 690 |
| Germany | 581 |
| Portugal | 470 |
| Ireland | 461 |
| Luxembourg | 421 |
| Spain | 302 |
| Greece | 231 |

SOURCE: Investigación y Ciencia, December 1992, and elaboration of our own.

terns seems a necessity, although the direction of change is not clear.

The EC Commission is aware of the need and has undertaken some steps to follow in higher education and professional training, but, at the same time, it has been looking to adapt the diverse educational systems into a common ground in order to facilitate the free circulation of skilled personnel.

The economic implications of educational policies are also being recognized. Big, multinational enterprises and owner-operated organizations have expressed their views on the need for a greater number of scientists, engineers, and technicians.

This demand resists the diminishing interest of young Europeans to study in the scientific and technical domains. Therefore, there exists a social debate about the main trend to be followed in the educational process as a whole and in relation to the apparent conflict between classical culture and that oriented toward empirical and practical goals. The debate has extended to secondary schools. In the developed western European countries, this training is strongly related to well-reputed professional skills and, therefore, asks for a separation of a more general primary school. A major part of the blooming of the manufacturing industry in the German and Dutch regions lies in their asset of a highly skilled human capital. The situation is yet different in southern European countries, in spite of their efforts during the last few years to correct it.

## Comments and Conclusions on Educational Issues Related to Human Capital

It seems that there are still important differences in the educational profiles of developed and less developed countries. The main differences stem from the resources earmarked, the qualifications of the teaching staff, the suitability of the facilities, the characteristics of the secondary school and its relationship with professional training, and the connection of education with the labor market. The schooling rates of young Europeans at two ages in the EC countries are shown in Figure 2.

It is worth mentioning that the differences remain current at the present time and are being distinguished by a situation of general crisis of public support to educational policies. The forecast is, therefore, negative, in the sense that the differences are likely to remain the same or get worse. The adjustment of this trend cannot stand by single declarations of intent alone, but requires a real strategic policy with the availability of resources and appropriate goals.

On the other hand, to afford adequate response to the great variety of pressing demands, the educational system needs an important effort of flexibility and versatility. This implies, as in many other areas of the public arena, a better bond between expanding the social and economic agents and the pervading regional and local responsibilities (decentralization).

It is, therefore, essential to foster the participation of these agents (owners, trade unions, and regional and urban authorities) in designing, launching, and steering the educational policies, mainly in professional training and higher education. One cannot forget the high rate of unemployment for the young labor force in the less developed regions of Europe. Spain, for instance, had 50 percent of its population unemployed in the mid-1980s. Although this rate decreased during the last half of the decade, the values are still very high.

Using a biological analogy, it seems to us very pertinent to favor this selection of the population by putting it in touch with the adequate environment (i.e., the community must engage students once they have finished their education). In other words, it is necessary for the socioeconomic forces to play a role in helping administrators define educational policies. Alternatively, it is not possible to tackle the problem of heterogeneity by having a recourse to a blind, common, general force, or mechanism, for selection.

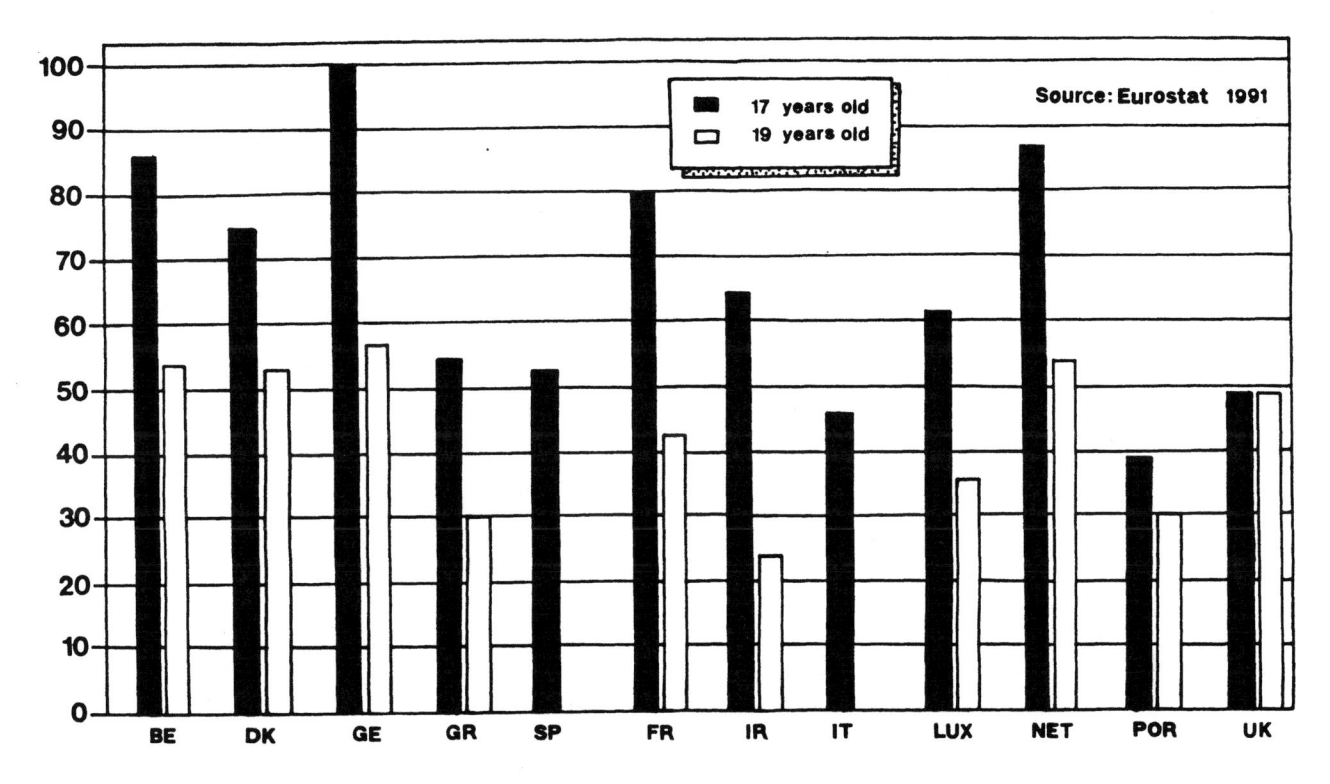

SOURCE: Investigación y Ciencia, December 1992.

**FIGURE 2** Schooling rates (in percent) of Europeans at two ages.

## EMPLOYMENT

Similar considerations can be applied to the employment situation. The evolution of the labor force in developed countries is characterized by the following trends:

- A substantial increase in the percentage of highly skilled and white collar personnel. (In some countries these professionals account for 60 percent of the active population.)

- A decrease, or stagnation, in the percentage of workers addressed to the primary or tertiary sectors (farming and services).

- A remarkable reduction in the percentage of manual workers (not related to agriculture), showing figures below 30 percent.

Therefore, developed societies appear to be, to a great extent, middle-class societies in which there is a continuous decrease in the number of manual workers in industry and services.

However, it is not clear whether the evolution from the less developed condition to the developed one is going to follow a gradual pattern or experience more abrupt changes derived from the availability of new technologies. The latter is already being seen in developing countries like those in Asia.

So, the data indicate that in the evolution process there is no substantial increase in the percentage of industrial workers. It is noteworthy to highlight the rise in the number of qualified professionals, administrative personnel, traders, and dealers.

This pattern denotes an evolutionary trend from developing countries to that of western developed countries before reaching a labor force profile of industrialized countries.

Again, this raises the conflict of world globalization with the singularities of every region. In this conflict, the market influences the international situation through the forceful behavior of multinational firms.

The wrestling of this mode of action with regional specifications and disequilibria is a key issue of reflection that pervades this paper.

## REGIONAL ANALYSIS AND SOCIOECONOMIC PARAMETERS

In this section, we will overview some macrostructural variables that may influence innovation in EC Objective 1 Regions.

### Demographic Trends and Labor Force

As a whole, the European Community's demographic evolution of the 1990s will remain stable at a figure of some 325 million people. For the period 2000-2015, an estimated decline of 1-4 percent is expected. With respect to the 1990 figure, in absolute figures, a decrease of 12 million people is foreseen.

However, differences among regions are, and will likely be for a long time, outstanding. During the present decade, the recessive demographic trend will consolidate firmly in most northern countries—Germany, Luxembourg, Belgium, and Denmark. It is presumed that prospective new member states of the EC will belong to this category. For three southern countries—Greece, Italy, and Portugal—a major change in population figures is not expected until the year 2000. On the other hand, France, the Netherlands, Spain, and the United Kingdom will likely undergo a moderate population increase, reaching demographic stability in the year 2000, whereas later evolution will invert this trend back to the 1990 figures in the year 2015. Ireland is expected to have a positive demographic growth at least until 2015.

Demographic trends will have a marked influence on labor markets. The EC labor force, as a whole, will slightly increase by 0.1 percent during the 1990s. This figure conceals the quick increase in many of the less favored regions—Ireland, Spain, Portugal, and Southern Italy—with a much more moderate growth, even a decrease, in the labor force in other member states. This does not, however, take into account the impact of immigration from countries outside the community. In absolute figures the increase in the community's labor force, based on its own demographic evolution, will only rise to 1.5 million people. This results from a 2 million increase in southern regions and a 0.5 million decrease in other areas of the community.

Capital intensive and highly profitable technological innovation, usually labor saving, has a broad market in northern countries, while southern countries will likely find difficulties in attracting investment for the creation of labor intensive businesses and/or economic activities, with massive jobs creating spillover factors (i.e., the car industry during the 1980s played this role in some less favored regions and seems to have exhausted its potential to go on with their positive activity).

## ECONOMIC INDICATORS

### Human Resources

From 1984 to 1990, employment in the community grew at an annual rate of 1.25 percent, resulting in a net increase of almost 9.5 million jobs. Each member state registered an increase in employment between 1985 and 1990, with great variability among them. For instance, the less favored regions in southern Spain and Portugal, to a lesser extent, enjoyed important increases in their employment figures.

By sectors, the 1980s saw the continuation of former trends of employment drift from industry to tertiary activities (services). From 1983 to 1988, the employment percentage in the tertiary sector in the community rose from 55 to 59 percent, while industrial employment decreased from 35 to 33 percent.

Although the community passed through a phase of positive employment growth, it is not at all clear if this growth was evenly distributed at the regional (NUTS II) level to be of any help in reducing the differences in unemployment and activity rates between more and less favored regions of the EC. Data show that the EC unemployment rate has decreased since 1986 but still ranked at about 8.3 percent in 1990. Among the factors that hinder job creation, especially in southern countries, are positive demographic growth, increased access of women to the labor market, and cyclic variation in activity, indicators combined with secular trends in the reduction of nonspecialized jobs. Concerning NUTS II level, differences are again substantial. On the one hand, there are 12 central regions with unemployment rates of less than 3 percent, while on the other hand 19 regions registered unemployment rates higher than 15 percent. (All figures refer to 1990.) Italy provides a special case. In 1990, it had the greatest inner regional disparities among all member states; some northern and central regions, such as Emilia-Romagna and Lombardy, showed unemployment rates less than 5 percent while most of the Mezzogiorno region suffered rates higher than 20 percent.

## Disparities in Productivity and Income

Differences among the EC regions, with respect to income per capita, is very acute. In 1988 the income per capita of the 10 most advanced regions was 3 times higher than that of the 10 less favored regions. These differences have remained stable since 1986 even though less favored countries achieved rates of economic growth higher than the community average and raised expectations of economic convergence. For instance, in 1986-1987, Spain, Ireland, and Portugal began a gradual movement of convergence toward the average EC gross domestic product (GDP) per capita, whereas Greek GDP per capita worsened. At the regional level, the average figure for the 25 less favored regions increased slightly with respect to the average EC GDP per capita. Nevertheless, there has been no progress among 10 less favored regions (mainly in Greece and Portugal) since the mid-1980s.

Besides these disparities, there are others with regard to productivity. In 1984 a slight trend of reduction of these disparities began among member states due to the improvement of the relative productivity in Portugal and Ireland. However, this trend stopped after 1987.

## Educational Opportunities and Training Facilities

Generally speaking, income per capita and unemployment differences reflect inter-regional disparities in productivity and competitiveness. Less productive and competitive regions face problems of generating higher and more evenly distributed income and of job creation. A lack of local infrastructures, mainly suitable in regions with more central and advanced ones, and a lack of a skilled workforce are crucial factors that hinder advances in economic competitiveness.

The problem of a lack of a skilled labor force is common to all regions within the EC. However, the causes differ according to their degree of development. In the more favored regions, fair economic conditions during the last decade have resulted in an imbalance between the competencies (and costs) demanded by firms and what the educational systems were able to supply. An additional problem is the re-skilling needed by an aged labor force in declining industrial areas. In less favored regions the lack of skilled personnel is due to the absence of modern and appropriate education and training facilities; the lack of cooperation of potential employers; rigid and outdated curricula; the lack of specialized and updated teaching staff; a very low degree of participation of active economic agents (union, firms, regional economic boards, etc.) in their planning and monitoring; and the absence of prospective and evaluative programs that are concerned about the adjustment between human resources supply and demand, work quality, and the encouragement of low level innovation activities. Lack of entrepreneurial skills and the absence of up-skilling training opportunities for the employed and unemployed are also problems widely present in these regions.

# SCIENTIFIC AND TECHNOLOGICAL DEVELOPMENT

From the current views of economic policy analysis, it seems evident that more and more innovation and investment in science and technology are key components for improving the competitive positions of societies.

The idea that technology is an endogenous product for economic wealth, as well as the suspicion that a science-technology system acts as a feedback mechanism fostering social and economic progress far beyond its initial value, is gaining strong support.

## Measurement of Science and Technology

Science, technology, and technological innovations are all abstract concepts that cannot be measured in a direct way. Although it might be interesting from a global point of view to treat science and technology activity as a defined subunit of economic performance, it is indeed difficult to correlate the factors influencing science and technology as inputs with economic outputs.

The most conventional indicators used currently, as both an input measurement of efforts devoted to research and development (R&D) and a part of the science and technology system, are those measuring economic and human resources. They are usually expressed as a percentage of GDP allotted to these activities and as a number of full-time equivalent researchers per unit of active population.

In-depth analysis reveals limitations in the applicability of these indicators with regard to specificities and subtleties of microenvironments in order to afford policies from a disaggregated point of view.

On the one hand, these indicators are too broad to allow the detection of the quality of the political measures undertaken on their basis. On the other hand, in spite of the international efforts made by OECD and Eurostat for homogenization and comprehensive accountability of R&D statistics, there is an evident absence of homogeneity in the information supplied by countries, leading to gaps in the collected data. All these difficulties are absolutely relevant for the purpose of this study, and so pervade it.

Therefore, one of the conclusions of this study will ask for better homogenization of indicators in terms of regional distribution and for an improvement in the type of indicators measuring human resources, both in absolute terms and in connection with education and employment. This information may allow the identification of the real potential in science and technology in less developed countries and regions, thus providing better insight for a comparison of these potentials with those of developed countries.

By doing so, we are following similar demands of science policymakers who are increasingly asking for new science and technology indicators (J. De Mother, *Science and Public Policy*, Vol. 19, p. 401-406, 1992). The debate on R&D indicators is a live question as shown by recent workshops of organizations such as OECD and Eurostat, or by the recent publication of a special issue of *Science and Public Policy* (Vol. 19, Numbers 5 and 6, October and December 1992).

In other words, we deem essential the disposition of data measuring disequilibria, as well as the relative strengths and weaknesses needed to promote political measures for science and technology other than those traditionally applied from the developed countries. A similar conclusion has been drawn by others. (See Padzerka as cited by J. A. D. Holbrook in *Science and Public Policy*, Vol. 19, Number 5, p. 266, 1992.)

### Economic Resources Devoted to R&D

The data available on R&D expenditure for the EC-12 less developed countries shows an uneven situation. For instance, Ireland is devoting between 0.8-0.9 percent of its GDP to R&D as a continuous trend. Italy is allocating more than 1 percent to these activities, whereas Spain, Portugal, and Greece have traditionally lagged behind. However, these three countries are making efforts to correct this trend at different rates. (See Figure 3.)

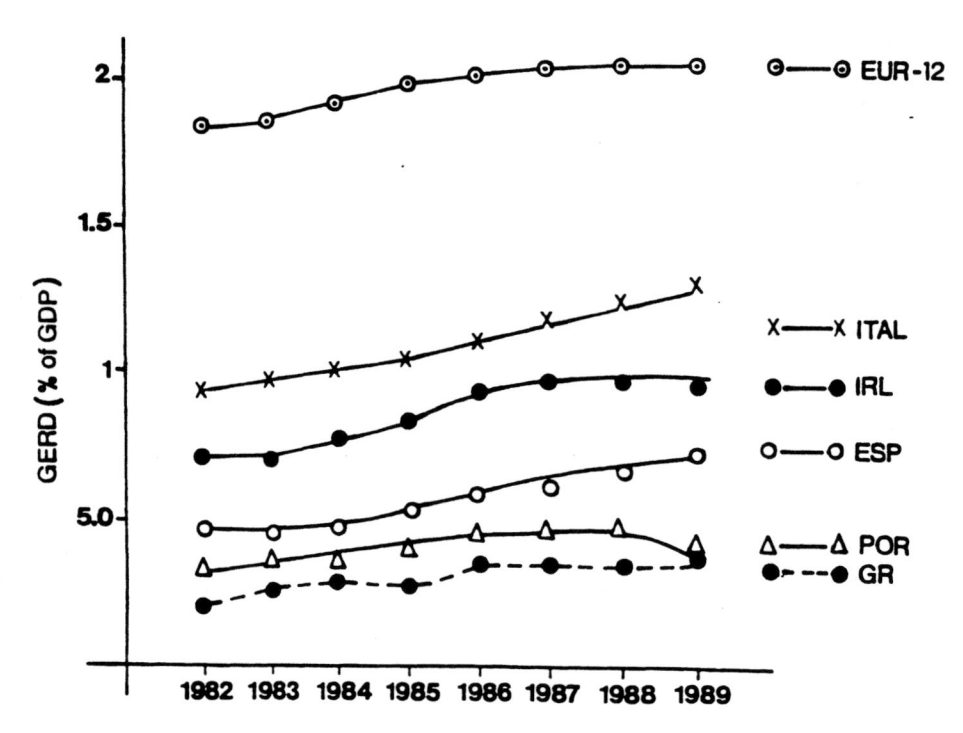

**FIGURE 3** Evolution of R&D expenditure in the EC less favored countries.

**TABLE 2** Regional Distribution of Expenditure in R&D Activities in Relation to Regional Gross Product, by Actors in Spain, 1988

| | GDP | | | | Intramural Expenditure in R&D by Sectors (millions of Pesetas) | | | | | | | |
|---|---|---|---|---|---|---|---|---|---|---|---|---|
| | | | Total Actors | | Enterprises | | Public Admin. | | University | | NPI | |
| Regional Communities | Total Thousands of millions | Percentage | Million | Ratio/GDP | Million | Ratio/GDP | Million | Ratio/GDP | Million | Ratio/GDP | Million | Ratio/GDP |
| Andalusia | 5,018 | 13.4 | 21,672 | 0.43 | 8,147 | 0.16 | 6,561 | 0.13 | 6,964 | 0.14 | | |
| Aragon | 1,250 | 3.3 | 6,938 | 0.56 | 2,684 | 0.21 | 2,027 | 0.16 | 2,222 | 0.18 | 5 | 0.00 |
| Asturias | 980 | 2.6 | 4,380 | 0.45 | 1,964 | 0.20 | 961 | 0.10 | 1,455 | 0.15 | | |
| Balearics | 879 | 2.4 | 824 | 0.09 | 201 | 0.02 | 148 | 0.02 | 373 | 0.04 | 102 | 0.01 |
| Canaries | 1,463 | 3.9 | 2,591 | 0.18 | 46 | 0.00 | 1,073 | 0.07 | 1,451 | 0.10 | 21 | 0.00 |
| Cantabria | 484 | 1.3 | 2,471 | 0.51 | 933 | 0.19 | 543 | 0.11 | 957 | 0.20 | 38 | 0.01 |
| Castile and Leon | 2,266 | 6.1 | 11,548 | 0.51 | 7,316 | 0.32 | 726 | 0.03 | 3,488 | 0.15 | 18 | 0.00 |
| Castile and La Mancha | 1,449 | 3.9 | 2,142 | 0.15 | 1,516 | 0.10 | 390 | 0.03 | 236 | 0.02 | | |
| Catalonia | 7,033 | 18.8 | 55,565 | 0.79 | 40,501 | 0.58 | 5,831 | 0.08 | 8,632 | 0.12 | 601 | 0.01 |
| Comm. of Valencia | 3,712 | 9.9 | 12,469 | 0.34 | 6,008 | 0.17 | 1,856 | 0.05 | 4,331 | 0.12 | 274 | 0.01 |
| Extremadura | 685 | 1.8 | 2,152 | 0.31 | 457 | 0.07 | 1,023 | 0.15 | 672 | 0.10 | | |
| Galicia | 2,131 | 5.7 | 6,065 | 0.28 | 1,953 | 0.09 | 2,085 | 0.10 | 2,027 | 0.10 | | |
| Madrid | 5,614 | 15.0 | 120,661 | 2.15 | 67,471 | 1.20 | 40,968 | 0.73 | 11,511 | 0.21 | 711 | 0.01 |
| Murcia | 952 | 2.5 | 3,675 | 0.39 | 1,191 | 0.13 | 1,455 | 0.15 | 1,026 | 0.11 | 3 | 0.00 |
| Navarre | 618 | 1.7 | 2,310 | 0.37 | 2,112 | 0.34 | 198 | 0.03 | | | | |
| Basque Country | 2,545 | 6.8 | 24,452 | 0.96 | 20,663 | 0.81 | 723 | 0.03 | 3,069 | 0.12 | 27 | 0.00 |
| La Rioja | 322 | 0.9 | 354 | 0.11 | 237 | 0.07 | 117 | 0.04 | | | | |
| Ceuta and Melilla | 95 | 0.3 | | | | | | | | | | |
| No Regional * | | | 7,419 | | | | | | 6,952 | | 467 | |
| SPAIN | 37,401 | 100.0 | 287,688 | 0.77 | 163,370 | 0.44 | 66,685 | 0.18 | 55,366 | 0.15 | 2,267 | 0.01 |

* Includes some universities: UNED (Open University), Private University of Navarre and Fellowships; and NPI: Grants to individuals.

SOURCE: R&D Statistics, 1988. Regional Accountability of Spain, 1985 Base, 1985-1988 Series.

This path of heterogeneity is even more acute when the data on R&D expenditure are presented from a regional perspective (as it is now possible to do for Spain and Italy). Table 2 shows the data corresponding to the regional distribution of 1988 R&D expenditure in Spain. One region, Madrid, possesses a level convergent with that of the EC average (2 percent of GDP), while none of the other regions reach 1 percent.

The inequitable model is also sustained by Italy but with a distinct shape. Nearly 92 percent of the total national expenditure in 1987 was concentrated in northern and central Italy (representing about 64 percent of the Italian population). The remaining 8 percent was spent by southern Italy, with 36 percent of the demographic force. (See Table 3.) On the other side, the regional distribution in this meridional part of Italy presents a more uneven distribution. (See Table 4.)

Data available for Portugal points to the same direction, since about 70 percent of R&D expenditure takes place in the region near Lisbon.

## Human Resources in Science and Technology

Human resources are the most important component of any national research system. As a logical consequence, personnel data represent the second main input indicator of R&D activities. However, its incorporation to this pertinent position in the R&D panorama occurred later than that of economic resources and was the result of the credit gained by the model proposed by H. Brooks on science planning and priority setting. This model viewed research not only as a technical overhead but also as a social overhead investment in which research and education, in broad terms, are inseparable activities and must be intertwined.

Since this incorporation, the measurement of human skills in connection to R&D has not been an easy task. Seminal to this issue was the Frascati manual as a highly valuable sample of OECD care, interest, and effort for standardization of R&D data. However, it seems obvious that in spite of the recognized merit of Frascati, it offers, today, limitations because of the increasing complexity of the relationship between science, technology, economy, and society. The manual cannot be taken as a made-to-measure suit for any model other than science-technology systems.

The situation has already been recognized by the main international organizations (UNESCO, OECD,

**TABLE 3** Total Expenditure in R&D Public Sector and Undertakings by Main Regions in Italy, 1987

| Region | Expenditure (millions of lire) | Percentage |
|---|---|---|
| Southern Italy | 6,238,983 | 66.84 |
| Central Italy | 2,383,665 | 25.54 |
| Northern Italy | 711,285 | 7.62 |
| TOTAL | 9,333,933 | 100.00 |

SOURCE: Original ISTAT, 1990, taken from Science and Technology in a Marginal Area: The Case of the Italian Mezzogiorno, R. Zobbi, December 1991.

**TABLE 4** Regional Distribution of Total Expenditure in R&D, Public Sector and Undertakings in Italy, 1987

| Regions | Absolute Expenditure | Percentage |
|---|---|---|
| Northern Italy | 6,238,983 | 66.84 |
| Piemonta | 2,002,535 | 21.45 |
| Valla d'Aosta | 6,821 | 0.07 |
| Lombardía | 2,635,464 | 28.24 |
| Trentino A.A. | 35,087 | 0.38 |
| Veneto | 346,435 | 3.71 |
| Friuli V.G. | 139,142 | 1.49 |
| Liguria | 463,581 | 4.97 |
| Emilia Rom. | 609,918 | 6.53 |
| - IT. Nort/occ. | 5,108,401 | 54.73 |
| - Nort/or. | 1,130,582 | 12.11 |
| Central Italy | 2,383,665 | 25.54 |
| Toscana | 431,354 | 4.62 |
| Umbria | 24,312 | 0.26 |
| Marche | 38,721 | 0.41 |
| Lazio | 1,889,278 | 20.24 |
| Southern Italy | 711,285 | 7.62 |
| Abruzzi | 63,252 | 0.68 |
| Molise | 512 | 0.01 |
| Campania | 347,229 | 3.72 |
| Puglia | 89,789 | 0.96 |
| Basilicata | 37,926 | 0.41 |
| Calabria | 17,602 | 0.19 |
| Sicilia | 1,151,625 | 1.23 |
| Sardegna | 39,813 | 0.43 |
| TOTAL | 9,333,933 | 100.00 |

SOURCE: Original ISTAT, 1990, taken from Science and Technology in a Marginal Area: The Case of the Italian Mezzogiorno, R. Zobbi, December 1991.

and the ECC) involved in the measurement and understanding of trends in R&D. This has led to the lively debate that is currently taking place. Some of the issues that have sprung up are as follows:

1. It is deemed necessary to discuss science and technology in a broader concept than that of R&D in current literature.

2. It is admissible to talk about human resources in science and technology instead of R&D personnel.

3. It is recognized that the debate should delve into the activities related to science and technology, enlarging the focus from those *related to science and technology production* to others indispensable *for a science and technology policy based on dissemination*.

4. Against the Anglo-American view, the humanities are kept within the frame of activities and disciplines involved in science and technology.

5. The difficulties in making adequate comparison between professional skills, both by qualification and occupation, must not exclude accountants and technicians, whose jobs are fundamental to rational and well-structured development of science and technology goals and projects.

6. In any case, the approach looking for a better link between R&D and education may take into account 5, 6, and 7 ISCED levels, as well as consider the distinction of doctoral studies as an independent level.

7. The idea of counting personnel as total numbers, as heads, without relying exclusively on counts of full-time equivalents is gaining intensity.

8. In a similar context it is considered appropriate to combine the measure of the number of people available (stock) with the concept of mobility, or flows, of personnel.

9. The human resources involved in science and technology have to be considered both as inputs and outputs to the system.

10. There is an evident conflict between the classifications concerned with skilled personnel by both qualification (education) and occupation. The next generation will see the continuing debate about this matter, focusing on the count of students (graduate, doctorate, and postdoc).

All these topics appear highly relevant to the present study, and their concerns and problems influence it.

### Data on Human Resources in Less Favored Regions of EC-12

The data currently available confirm the trends outlined before: (1) an absence of homogeneity in the information, and (2) the emergence of apparent disparities between countries and regions (see Tables 5-7). As can be seen, the information up till now contains data, although not complete, on stocks but not on flows. In neither is there too much detail in a breakdown by academic skills or degrees.

In order to correct the situation, we have prepared a survey/questionnaire addressed to heads of academic units, heads of R&D units, and personnel of companies that belong to EC Objective 1 Regions. This survey is now under way. As an illustration of its purpose, we include the introduction of the questionnaire:

> The aim of this questionnaire is to obtain first-hand, updated information about human resources devoted to scientific or technological research in institutional facilities located in the less favored regions of the EC. It is expected that the data retrieved by these means can be helpful in designing EC science and technology policies more appropriate to the needs of these regions. Data on personnel mobility are particularly appreciated due to the importance attached to this information as indicator of a research unit situation and as a probable area of political and financial action. For the sake of this study we define *mobility* as any variation in the number and/or composition of the human resources of a research unit. Most careful attention and accuracy in the fulfillment of this questionnaire is earnestly requested. Thank you for your kind cooperation.

**TABLE 5** Summary of the Collected Information about Personnel from R&D Activities in the Less Favored Countries of EC-12

| Occupation Sector Countries | Administration FTE | HC | Higher Education FTE | HC | Post-Graduate Grant | All | Enterprise FTE | HC |
|---|---|---|---|---|---|---|---|---|
| Greece | X | - | X | X | X(CC) | - | X | - |
| Spain | X | X | X | X | X(CC) | - | X | X |
| Ireland | X | - | X | - | X(FTE) | X(CC) | X | - |
| Italy | X | X | X | X | - | - | X | X |
| Portugal | X | X | X | X | - | - | X | X |

Key:  x  denotes information available  -  denotes absence of information
   FTE  full-time equivalent        HC total number (head count)
   CC  conversion co-efficient

**TABLE 6** Summary of the Information Contained in the Comparative Analysis of the Science and Technology Policy of the Countries with the Less Favored Regions of Europe (EC-12)

| | | COUNTRY (Year of Last Information) | ITALY (1988-89) | IRELAND (1984) | SPAIN (1988-90) | PORTUGAL (1988-90) | GREECE (1984-87) |
|---|---|---|---|---|---|---|---|
| R&D Indicators | Inputs | Expenditure (GERD) | + | + | + | + | + |
| | | - By Sector | + | + | + | + | + |
| | | Personnel (n² per 1000 active) | + | +/- | + | +/- | + |
| | | - By Sector | + | - | + | + | + |
| | | - By | + | +/- | - | + | + |
| | | - By Field of Science | +/- | + | + | - | +/- |
| | Outputs | Number Publications | - | - | + | - | - |
| | | Patents | - | - | + | - | - |
| Analysis of Special Sectors | University | Human Resources | - | - | + | - | - |
| | | - Distribution by Field | - | - | + | - | - |
| | | Students | - | + | + | - | - |
| | | - Distribution | - | + | + | - | - |
| | | Research Training Grants | - | - | + | + | - |
| | | - Distribution | - | - | + | + | - |
| | Enterprises | Expenditure | - | + | + | - | - |
| | | - by NAABS | - | + | + | - | - |
| | | Personnel | - | - | - | - | - |
| | | Distribution (public, private, SME) | - | - | +/- | - | - |
| | | | - | - | +/- | - | - |
| | Regional Distribution | Expenditure | +/- | + | + | +/- | +/- |
| | | Personnel | +/- | + | + | | |

Key:  +  denotes complete information
   +/-  denotes incomplete information
   -  no information

SOURCE: Copol 1990, Tendances de la recherche scientifique et du developpement technolgique dans la CEE; Trends Research and Technological Development in the ECC (Rapport/Report EUR 13795, FR/EN); Greece, Copol 88, Report EUR 11983 EN; Portugal, Copol 91, Rapport EUR 14343 FR; Italy, Copol 90, Report EUR 13313 EN Ireland, Copol 88, Rapport EUR 11980 FR; Spain, Copol, Report prepared by Mr. J. Elorrieta Jove, EN.

**TABLE 7** Summary of the Situation of Personnel Devoted to R&D in the Countries with the Less Favored Regions of Europe (EC-12)

| COUNTRY | | RESEARCHERS | | TECHNICIANS | | OTHERS | | TOTAL | |
|---|---|---|---|---|---|---|---|---|---|
| | | Number | % | Number | % | Number | % | | |
| ITALY | Total | 67,844. | 100.0 | 32,892. | 100.0 | 21,616. | 100.0 | 122,353. | |
| | Public Sector | 35,279.[1] | 52.0 | 12,006.[2] | 36.5 | 8,798.[2] | 40.7 | 56,083.[2] | (64,637[2]) |
| | Enterprises | 32,565.[1] | 48.0 | 20,886.[2] | 63.5 | 12,818.[2] | 59.3 | 66,269.[2] | (57,715[2]) |
| PORTUGAL | Total | 5,003.6 | 100.0 | 3,571.5 | 100.0 | 2,308.3 | 100.0 | 10,883.4 | |
| | Public Sector | 4,237.7 | 84.7 | 2,385. | 66.8 | 1,673.9 | 72.5 | 8,296.6 | |
| | Enterprises (+ NPO) | 765.9 | 15.3 | 1,186.5 | 33.2 | 634.4 | 27.5 | 2,586.8 | |
| GREECE (1983-87) | Total | 4,000. | 100.0 | 1,572. | 100.0 | 1,908. | 100.0 | 7,480. | |
| | Public Sector | 3,162. | 79.1 | 1,278. | 81.3 | 1,535. | 80.5 | 5,975. | |
| | Enterprises | 838. | 20.9 | 294.[2] | 18.7 | 373.[2] | 19.5 | 1,505. | |
| IRELAND (1984) | Total | 3,626. | | 1,336. | | 1,231. | | 6,193. | |
| SPAIN (1988) | Total | 59,592 | 100.0 | | | | | | |
| | Public Sector | 48,623 | 81.6 | | | | | | |
| | Enterprises | 10,069 | 18.4 | | | | | | |

Footnotes:
[1] Estimated from breakdown by type of research.
[2] It appears as such in breakdown by sector of activity.

SOURCE: Copol 1990, Tendances de la recherche scientifique et du developpement technolgique dans la CEE; Trends in Scientific Research and Technological Development in the ECC (Rapport/Report EUR 13795, FR/EN); Greece, Copol 88, Report EUR 11983 EN; Portugal, Copol 91, Rapport EUR 14343 FR; Italy, Copol 90, Report EUR 13313 EN; Ireland, Copol 88, Rapport EUR 11980 FR; Spain, Copol, Report prepared by Mr. J. Elorrieta Jove, EN.

This questionnaire is divided into five short sections. The first one is thought to reveal the composition and characteristics of all the personnel making up the human resources of scientists, technologists, technicians, and other researchers (fellows, grant holders, etc.) working in the research unit under survey. Gender, age, position, activity, and level of training features of the researchers are requested.

The second section of this questionnaire is aimed to provide an image of the mobility resulting in the present composition of the human resources of the research unit in this survey. Geographical and educational features are requested in order to single out those people proceeding from non-Objective 1 Regions, and then data on gender, age, position, activity, cognitive area, and level of training are asked for these people.

The third section of this questionnaire points to the mobility of scientists (commonly people who graduated from Science Schools doing R&D; what the Frascati manual calls "researchers"), technologists (commonly people who graduated from engineering schools and who deal with technical activities of R&D or tasks of adaptation, diffusion, troubleshooting, quality control, routine tests, etc.), technicians and equivalent staff, and other researchers (fellows, grant holders, etc.) during 1992. Temporal, geographical, organizational, and motivational features of the researchers undergoing mobility are requested.

The fourth section of this questionnaire asks for a few characteristics of the research unit under survey. Specific questions are asked on area of activity, sources of support, and the allocation of financial and human resources.

Finally, for technical reasons—and always respecting the confidentiality of respondents—we need some identification data. (VERY IMPORTANT NOTICE: Less favored regions are named Objective 1 Regions in the terminology of the EC. This survey is addressed to research units in this kind of region. Although there are a few others within the EC, the Objective 1 Regions considered in this survey are the following: Greece, Ireland, and Portugal as whole countries; Abruzzi, Basilicata, Calabria, Campania, Molise, Puglia, Sardegna, and Sicilia within Italy; and Andalucía, Asturias, Islas Canarias, Castilla-León, Castilla-La Marcha, Comunidad Valenciana, Extremadura, Galicia, and Murcia within Spain. It is crucial to take into account these classifications in order to answer correctly the following questions on geographical mobility of the researchers!)

## SOME CONCLUSIONS AND RECOMMENDATIONS FROM THIS ANALYSIS

- The study of human capital in science and technology needs more work from the statistical approach. All initiatives that are being undertaken by different organizations are welcome, although they are likely to require a serious discussion from a multidisciplinary point of view. Mobility is for instance fundamental to understanding the problems of less favored regions.

- The educational programs related to development of skills for science and technology must be discussed with the social agents. A partnership between owners, trade unions, and public administrators appears to be essential to establish the great lines of training needs.

- In order to foster the development of less favored regions through science and technology, an appropriate selection of priorities is fundamental. For this selection it is necessary to take into account the inner characteristics (its strengths and weaknesses) of every region. Adoption or mimicking of too general priorities is most likely hopeless.

- It is fundamental to favor the creation of an adequate infrastructure to permit the incorporation of skilled personnel. An appropriate general environment may encourage this trend.

- It is perhaps time for a plea for more creativity and flexibility of policies concerning human resources for science and technology. Since heterogeneity appears as the main characteristic, specific measures and programs should be applied.

## SOME PRELIMINARY CONCLUSIONS AND COMMENTS FROM THE INQUIRY CARRIED OUT BY OUR GROUP

The response from the 3,000 units surveyed by mail has been reasonably good (about 30 percent), taking into account the mailing characteristics of the survey, the complexity of the questionnaire, and the limitations of mail services in these EC 1 Objective Regions.

The complete information from the survey will be communicated to the EC Directorate General XII from which we are contractors.

Nevertheless, a few conclusions and comments can be drawn at this stage:

- The information gathered confirms the difficulties in obtaining relevant data concerning human resources in science and technology. This general conclusion has to be stressed in view of the considerable extent of disaggregation concerning survey units reached by us. One has to wonder about the significance of data obtained from survey units with a greater degree of aggregation.

- One reason for this difficulty concerns the breakdown by qualification. This owes first to the evident heterogeneity existing among countries with regard to degrees and length requirements for their attainment. Second, it seems difficult to reconcile the extent and variety of educational degrees and diplomas with the ISCED levels (for instance, levels 5, 6, and 7, and the absence of a doctoral level). An effort for homogenization, at least in terms of equivalence analysis, seems indispensable and must be rewarding.

- Another side of the problem derives from the lack of information on human resource issues among R&D managers of whether they belong to academic institutions or undertakings. A specific action on the formation of these questions may be worthwhile.

- A third problem comes from the attempt to match the restricted R&D activities and their statistical data following the Frascati manual with those more broadly comprised under the heading science and technology.

- We have noticed that among academic or scientific institutions from Greece, Italy, Ireland, and Portugal there is a marked heterogeneity in the doctoral distribution from none to a reasonable level. However, there also seems to exist a potential for doctoral training inside the regions or even for attracting foreign doctors to them. This is a puzzling situation since I consider doctoral training as part of a learning by doing teaching process, a first and necessary step to improve the scientific environment in the less favored regions. Moreover, the procurement of a doctoral degree in the same region from each individual seems to me—paradoxical it may seem—a better onset for scientific careers until there is sufficient development of the discipline in that region. A better likeness to reality can follow from this path.

- The situation is not the same in Spain, probably as a result of the administrative requirement of a doctor's degree to pursue an academic career.

## NOTES

1. In collaboration with T. González de la Fe, J. M. Iranzo, R. Blanco, and C. E. García.

The study based on the questionnaire survey mentioned in the text was ended and approved in June 1993. It was carried out under Contract Number STRI-0020-ES from the Commission of European Communities (Directorate General XII, Science, Research, and Development). The Final Synthesis Report entitled *Study on Human Capital for S/T in the Five Less Favored Member States* is owned by that Directorate and may be available from it.

We are indebted to Mrs. Mercedes Tenjido and Mrs. María Jose Beltrán for their secretarial help.

# The Longitudinal Analysis of the Selection of Careers in Science and Technology

## Jon Miller

In this commentary, I want to discuss the rationale for longitudinal measurement in the study of career choice, review the strengths and weaknesses of the papers by Dr. Xie and Dr. Muñoz, and suggest some approaches to further research that improve our understanding of the career selection process. Having worked for several years in the conceptualization and implementation of a major longitudinal study in the United States, I am pleased that a panel focused on longitudinal studies was included in this international conference. The increasing number of longitudinal studies and the focus of serious academic and policy discussions on these studies marks a growing recognition of the value of this kind of work, and I am pleased to be a part of this process.

## THE IMPORTANCE OF LONGITUDINAL MEASUREMENT

It may be useful to begin with a brief discussion of the importance of longitudinal studies in thinking about the selection of careers throughout the life cycle and in testing our hypotheses and assessing the impact of our efforts to impact this process. I believe that the growing number of good longitudinal studies has had a beneficial impact on the ways that we think about the selection and achievement of careers and may even have enriched our theories about these processes.

When investigators had available only cross-sectional time series measurements, the picture of the career selection process was viewed as more linear, incremental, and rational. And many of the models were consistent with the time series data, showing, for example, a steady decline in student plans for a career in science and technology during the high school and college years. This observation may be applied to social science more generally, but it was undoubtedly true in regard to models of career choice.

With the access to longitudinal data sets that included career choice information, it has become apparent that the process is much less stable and far more dynamic than many of the more linear and static models would have predicted. The fundamental point is that studies of career choice are studies of human behavior. We are interested in changes in the plans, expectations, skills, and behaviors of individuals, and the data from good longitudinal studies provide the level and kind of measurement that supports the development and testing of models of human choice and change.

## XIE: A DEMOGRAPHIC MODEL

The Xie paper offers a demographic model of movement into and out of the science and technology career stream, using a modified Markov chain approach. The introduction of a demographic model makes the paper useful at the conceptual and heuristic levels. A series of misspecifications of central variables, the disjuncture of measurement concepts across longitudinal segments, and a 20-year gap in the two databases combine to erode the substantive value of the conclusions. Let me discuss each of these points in greater detail.

Xie's discussion of the movement of individuals into and out of the science and technology career stream is useful. It emphasizes the dynamic nature of the process and points to one important use of longitudinal studies. Conceptually, the ability to look at the transition probabilities across time and at different levels (and types) of schooling would be a useful macro-level tool to understand and monitor the pipeline of scientific and technical personnel. This kind of macro analysis should complement micro-level analyses of family, school, peer, media, and market influences on career choice, not replace it. As we know from economic analysis, macro-level theories built on inaccurate assumptions about micro-level behaviors seldom work.

Unfortunately, the elegance of the conceptual model is not paralleled in the empirical specification of the model. There are several serious problems.

First, Xie is vague about the definition of scientific and technical occupations. Since the analysis is limited to those students planning to enroll at the university level, it would appear that all sub-baccalaureate technical occupations are excluded. It is not clear, however, whether the health professions are included or excluded from the analysis, and that determination will have a significant impact on the subsequent analysis of gender differences in the selection of scientific and technical careers. Xie provides a footnote saying that the social sciences are excluded and offers to provide occupational codes upon request, but this is information that is essential to the reader in seeking to assess the proposed model.

Second, for the 7th through 12th grade segment of the longitudinal chain, Xie defines all students who "strongly agree" with the statement that they enjoy science as "belonging to the state of intended science/engineering postsecondary education." While it is reasonable to expect that most students who are thinking about a scientific or technical career would say that they enjoy science, it is very likely that a significant proportion of the students who strongly agreed with this statement have clear career intentions in fields other than science and technology. Some students report that they enjoy or like virtually all of the subjects that they study, while others tend to be more reserved and give all, or most, subjects a more moderate rating, including those related to possible career preferences.

The definition used for the student population, however, should bear some relationship to the definition used for the college population and to the definition of the final occupational set. For example, if medicine, nursing, and other health professions are included in the definition of a science/engineering occupation, then it would be reasonable to use a somewhat broader definition. If, on the other hand, the final definition is the National Science Foundation's classification of graduate-educated scientists and baccalaureate-educated engineers, then a broad definition at the middle and high school levels would lead to the false conclusion that large numbers of students who always intended to be physicians had left the pipeline, when, in fact, they had never been in it. Given the large number of more specific and career related questions included in the Longitudinal Study of American Youth (LSAY) database, Xie could have operationalized this variable in a more precise and relevant manner.

Third, a similar problem occurs with the definition used by Xie for the college years. Xie identifies a set of college majors as representing a career interest in science and engineering. While this is clearly a more precise definition than was used in the preuniversity years, it is equally vague as to boundaries and content. If the health professions are included, then majors may be a reasonable approximation. If the health professions are excluded, then we would again interpret all of the premedical majors in biology or chemistry as being in the science and engineering pipeline and subsequently dropping out of that pipeline, which would be an erroneous conclusion. Given Xie's later effort to make generalizations about gender differences, the issue of the inclusion or exclusion of the health professions is a central problem since, in the United States at least, a slight majority of high school students planning to enter medicine are female.

Fourth, in an effort to develop a synthetic cohort over a longer time span, Xie combined the precollege data from the LSAY and the college and postcollege data from the National Longitudinal Study of 1972. This linking might work in areas that have been relatively stable over the last two decades, but it does not work for the purposes of Xie's analysis. In the area of gender differences in course enrollment and career preferences, there have been major changes over the 19-year period between the two studies. For example, had Xie looked at student use of computers in this synthetic cohort, the results would have shown a major drop-off between the ages of 17 and 19. We know from numerous other studies that this would be a wrong conclusion, but the rate of computer usage was significantly lower in the early 1970s and the merger of these two data sets on substantive areas that are undergoing rapid change is inappropriate. During these two decades, the proportion of high school and college women who have been taking advanced courses in science and mathematics has increased significantly, as has the number of young women planning careers in science, mathematics, medicine, and engineering.

Finally, these problems of definition, specification, and time breaks combine to raise serious doubts about the substance of Xie's conclusions. Xie claims that the major loss of young women from the science and engineering pipeline occurs between the ages of 17 and 19, but the observation that a very similar drop occurs for males leads one to suspect that the real factor may be the change in definition from the enjoyment of science to enrollment in a specified set of college disciplines. It is very likely that some part of the observed drop (by any definition) is the result of comparing a high school cohort from the late 1980s with a college cohort from the early 1970s, disregarding two decades of growing feminist awareness and increasing rates of female participation in advanced science and mathematics courses.

## MUÑOZ: HUMAN RESOURCES IN LESS DEVELOPED REGIONS OF EUROPE

The Muñoz paper addresses both the conceptual problem of the linkage between human resources and job availability and the problem of insufficient data to fully understand the processes. It is useful to think of these two issues separately.

Muñoz documents that the less developed regions of Europe have fewer persons trained in science and technology (by almost any definition) and that the science and engineering pipeline is smaller in these regions than in the more developed regions of Europe. He observed lower rates of public spending for education and lower numbers of students seeking higher level education and training in science and technology-related occupations. There are similar regional disparities in the United States and in most industrial nations.

Muñoz also observes similar patterns in regard to public investment in research and development (R&D) and in the level of employment in occupations and institutions related to scientific research and technical development. Again, this pattern would be found in the United States and most industrialized countries. In recognition of this pattern in the United States, the National Science Foundation has created a special grant program to provide support for scientific work in selected states where universities and other research centers have not been able to compete successfully against similar institutions in other states.

The issue is the linkage between these two sets of observations and the implications for public policy. Historically, some economists and development specialists would argue that once there are jobs and opportunities available in a region, the labor market will adjust and more young people will begin to prepare for these jobs, seeking the kinds of training needed. This is a demand driven model. While it clearly works for some kinds of economic development, it is not clear that it will work equally well for scientific and technical development. Given the long period of training needed for many higher level scientific and engineering positions, it is more likely that skilled personnel from other regions will be drawn to new job opportunities in less developed regions, unless there is a general shortage of skilled personnel in all regions. Over a longer period of time, new generations of young people growing up in an area with available positions in science and engineering will begin to think about those opportunities, but the short-term change may be relatively small, especially at the professional level.

An alternative view has been that an increased supply of available skilled personnel in a region will make that region more attractive to potential employers, who may then elect to locate a new research or development facility in that region. In a market with a rapidly increasing R&D sector and a general shortage of skilled personnel, this model may work effectively, but those conditions are rarely met.

Muñoz, as many before him, struggles with the linkage between the supply of human resources and the level of economic activity in R&D. He recognizes that there is a linkage, but he does not suggest a causal order. I suspect the linkage is circular, with increasing job availability and increasing numbers of young people seeking careers in science and technology mutually reinforcing each other. It is clearly a dynamic model.

In his search for some structure to this problem, Muñoz finds a dearth of data that is both directly relevant and sufficiently precise. He notes the differences in definition of scientific and technical jobs and the difficulties of making valid comparisons among non-identical data sets. The work of the European Community in seeking more standardized data has produced some marked improvements to date, but, undoubtedly, more needs to be done. In addition, there are important kinds of data (like the longitudinal studies used by Xie) that are not being collected in the European Community at the present time.

## THE IMPACT OF THE END OF THE COLD WAR

The end of the Cold War will have an important impact on all of our models of the demand for and supply of scientific and technical personnel, and it is important to recognize this potential impact. In both the United States and the European Community, but especially in the U.S., a significant portion of scientific and technical personnel have been engaged in defense-related R&D. Given the relative stability of this sector of the R&D enterprise over the last four decades, we have been able to develop pipeline models of human resource preparation with some degree of certainty about the relative size of the demand for new graduates.

With the end of the Cold War, all of the western governments are moving substantial amounts of resources from military-focused spending to civilian-focused spending. In the United States, the Clinton administration has announced a major effort to convert many of these defense industries into nonmilitary production of consumer goods and environmental protection processes and goods. It is far too early to guess at the likely success of this conversion, but it is equally clear that we do not know what level of civilian R&D will be supported by our economies as presently structured.

Consider two possibilities. Assume that the movement of defense-related scientific and technical personnel into the civilian sector will take a decade to achieve, or that it will be a decade before we can determine the longer-term level of resources that western governments will continue to devote to military purposes. In the meantime, the personnel and facilities that are converted from defense to civilian purposes meet, or largely meet, the growth in R&D personnel needs of the civilian sector. The opportunities for new graduates would diminish and the pipeline flow would drop in rough proportion to the availability of positions. In this model, the demand for scientists and engineers would drop and the incentives to relocate to presently less developed regions would be greatly reduced.

The most optimistic model is that the flow of talent from military to civilian objectives will stimulate a new surge of innovation, creating new products, new industries, and new jobs. Surely we would all hope for this model, but realistically, it is more difficult to imagine. If, as a recent television show suggested, this new technology should lead to the development of electric automobiles that would cause less harm to the environment, there might be an increase in the design, manufacture, and repair of electric automobiles, but would there not be some decline in the level of economic activity associated with gasoline-powered automobiles? And what if electric automobiles turned out to require less time to manufacture or less frequent repairs? Overall, I would expect that there may be a net gain in economic activity in the civilian sector, but I would expect that it will take at least a few years to fully offset the decline in military-related economic activity. We stand at a very interesting point in time, and we should not assume that our previous models of supply and demand for scientific and technical personnel will continue to work in future decades without some major re-thinking.

## WHAT DO WE NEED TO KNOW ABOUT THE PROCESS?

I understand the purpose of this international conference to be a reporting and sharing of current research in regard to scientific and technical careers and the identification of future directions for this research. Let me conclude, therefore, with some brief observations about some substantive issues that require a new or substantially increased level of research attention.

First, I want to stress the importance of longitudinal measurement in the study of the career choice and attainment process. It is a dynamic process, and we need to build theories and models that reflect the vitality of the process. After more than two decades of working with both cross-sectional and longitudinal data sets, I am firmly convinced that we must have more and better longitudinal data sets if we wish to make substantial improvements in our understanding of these processes.

In the United States we have initiated a number of important national longitudinal studies, but there are still important misunderstandings about the nature of and need for these studies. Some policy leaders and agency personnel seem to think that one national longitudinal study every eight years is sufficient to answer every possible question about the educational process. Virtually none of these people has ever attempted to analyze a longitudinal data set or has ever worked to combat respondent fatigue over the years and sustain cooperation. The LSAY, which I conceptualized and of which I am the director, represents a new and important example of a kind of longitudinal study that needs to be replicated in other subject realms and in other countries.

In contrast with the National Longitudinal Study of 1972, the High School and Beyond Study, and the National Elementary Longitudinal Study of 1988, the LSAY focused on the development of student interest and achievement in a limited subject range—science and mathematics—and attempted to measure a wide range of factors that affected those outcomes. We tested each student in science and mathematics each year, collected student assessments of courses and reports of school activities each semester, interviewed one parent of each student once each year by telephone to learn more about the home environment and to obtain parental estimates of student time use and activities, collected reports from each science and mathematics teacher that served an LSAY student about the content of the course, and obtained school level reports from the principal of each school periodically. In the first 6 years, we collected over 6,000 items of information on each of about 7,000 students. It has been an intensive look, but it is the way that systemic measurement must be done if we are to develop a broader and more systemic understanding of how young people learn about science, mathematics, and technology.

In the future, we need to initiate new cohorts in the study of science and mathematics, and the government should support similar studies focused on how students acquire humanistic and social understanding, reading and language abilities, and political and social values. And it is essential that parallel longitudinal studies be developed in several countries at the same time, using common metrics. Unfortunately, even the best single national longitudinal study cannot measure some system level variables because most nations have a relatively common educational system. It is only when there are parallel longitudinal studies like the LSAY that we will be able to build models of student behavior that take into account both the family and school characteristics of each student and the systemic variables within which these other factors operate.

Longitudinal studies are expensive. It is the nature of the study, and it is unlikely to change. Pencils are less expensive than computers, and we still need some of them, but few people would seriously propose to substitute more pencils for computers for most purposes. Yet funding agencies in virtually every country still prefer to support a lot of smaller studies than a few larger ones. Perhaps the responsibility falls to those of us in the field to demonstrate the value of longitudinal data and to educate our students, who may become the next generation of agency administrators, about the need for longitudinal measurement.

Second, it is important to monitor the flow of young people into and out of the scientific and technical career stream. The work of Xie points to a direction that we need to pursue. While there will also be some flow into and out of this stream (and we should encourage students to change their minds if they are not happy with an earlier choice), we need to be able to identify the points in the stream where there are significant numbers of students exiting and to study the reasons why. Recently, Sheila Tobias has focused considerable attention on the impact of introductory college level science courses on the attitudes of students about science as a field of study and as a possible career. The kind of macro-level model proposed by Xie, with appropriate definitions and data, could be most useful in assessing points or difficulties that need further examination. A comparison of demographic, or flow, models in different countries or different regions—along the lines suggested by Muñoz—may be helpful in understanding more about the dynamics of economic development.

Finally, given the increasing length of life for many individuals, we need to think about new models that incorporate mid-life changes in career choice. While there has been some very useful discussion of mid-life changes in the educational and sociological literatures, there have been few databases suitable for use in testing major hypotheses or in constructing models of these behaviors. Again, and for all of the same reasons, we will need longitudinal measures of adults to be able to understand voluntary and forced modifications in career paths.

It is a large agenda and it will take resources, but it would be less expensive than devoting millions of dollars to job programs or economic development programs without an understanding of how those processes work. Neither governments nor economies will stand still while we study these issues, but we must seek to use our existing knowledge and our understanding of what we need to know to become more a part of the policy thinking processes.

# PART III

# Analyzing Trends in Science and Technology Careers: Factors Determining Choice

# Overview of Technical Papers

## Wendy Hansen

### BACKGROUND

Development and adoption of new technologies is key to economic prosperity in an increasing global economy, an economy that is altering economic and employment structures. As globalization develops and accelerates, our workforce will be a key determinant of our ability to compete and prosper.

Understanding the scientific and technical workforce is becoming a top priority of countries around the world. The increasing international mobility of our scientific and technically trained people adds additional and immediate pressures. It is now, more than ever, critical to understand just why individuals choose to pursue education and training in science and technology and to proceed on to careers in their specialized fields.

### INTRODUCTION

This summary paper has been prepared based on the papers submitted for discussion in Panel 3: Analyzing Trends in Science and Technology Careers: Factors Determining Choice. *Science and Technology Careers: Individual and Societal Factors Determining Choice* by Thomas Whiston and *Factors Behind Choice of Advanced Studies and Careers in Science and Technology* by Torsten Husén probe the factors influencing an individual's decision to pursue scientific and technical studies and to continue on to a career in science and technology.

### Factors Influencing Choice

What are the factors influencing a man or woman to pursue studies in science and technology? What factors contribute to the decision on whether or not to opt for a career in science and technology?

*1. Sociocultural Environment*

- Stereotyping manifests itself in childhood, shaped by the family environment, friends, and the community.

- A parent is a powerful role model. The parent may influence the choice of curriculum and scholastic achievement, directly and indirectly.

- Social acceptance—the status of science and technology both real and perceived—impacts upon the choice of the individual.

*2. Teaching of Science in Schools*

- The quality of teaching directly influences a student's scholastic achievements, limiting or enhancing program prospects. This can impact at an early stage when poor marks in mathematics and science limit future educational options, and also affect a student's self image and confidence in the subject. Poor teaching is a factor that not only may influence the individual's scholastic achievement, but also may inhibit the student's interest in science and his or her attitude toward science.

- The teacher provides a role model for the students. A good teacher can put the student at ease with science.

- The student-teacher relationship is an important factor in scholastic achievement and the student's future decisions.

- The workload of science and technology programs is often measured against that of other specializations and is often perceived as far more difficult and heavy.

- Science departments are often, whether real or perceived, separated from the rest of the school's programs and activities.

*3. The Gender Gap*

- Regardless of culture, men and women undergo different life experiences according to their gender. Choice of program and career are affected by gender stereotyping. For example, analyses show that there seem to be "male" and "female" specializations in science and technology. Men exhibit greater attraction to applied and physical sciences while women show greater propensity to life sciences like agricultural and biological sciences.

- Gender stereotyping may also influence scholastic achievement, as well as program choice. Young boys are encouraged to enter and excel in science; do young girls receive the same encouragement?

- Science and technology programs are seen as rather confined avenues of learning. Women may be deterred by the narrowness of learning that science programs offer.

- There is a perception of a hostility of the scientific community to accept women.

- Women are underrepresented in the scientific community—there are few role models to encourage women to pursue studies in science and technology and provide examples of successful careers in the scientific community.

- The academic and career path is seen as inflexible to women—reentry is difficult and women are led to feel they must choose between a career and a family.

*4. The Image of Science*

- Is science beneficial or harmful? Scientific contributions to society often seem unheralded while, with the help of the media, science is often perceived as harmful, even evil. It is seen as an ally to industries, which are themselves perceived as dirty and heartless.

- Science continues to have a mystique, making it rather unapproachable. It is often seen as an elitist discipline. Whether real or perceived, this may deter individuals from entering studies or careers in science.

*5. Career Prospects*

- Science and technology curricula are perceived as narrow, and thereby limiting career flexibility.

- The health of the economy influences the availability and attractiveness of jobs in science and technology. The perception of availability of good jobs and fiscal reward is an important factor in an individual's choice of program of study, as well as continuance in the field.

- The nature and type of scientific and technical activities change over time. Science is known for so-called "hot areas," and the built-in time lag of the educational process may lead to a mismatch of available talent and demand, and result in low demand in particular areas. This may in turn lead to perceptions of lack of jobs in science in general.

## IMPLICATIONS

An individual makes choices based on life experiences and perceptions of consequences of those choices. More often than not, decisions are based not on a single causal factor but on a series of events with various intertwined factors exerting their influence along the decision path.

Our questions still outnumber our answers. More research is needed to address the challenges nations around the world are facing—when and how to best intervene in the process to maximize influencing the educational and career choice of individuals to ensure we have a highly skilled scientific and technical workforce to face the challenges of today and bring us into the future.

# Science And Technology Careers: Individual And Societal Factors Determining Choice

## Thomas Whiston

### PROLOGUE: CAREER SELECTION IN THE SCIENCE AND TECHNOLOGY REALM

First, let me honestly state that this is a difficult topic to write about. For sure it is not that difficult to put forward (as I later do) a sequential dependency model that attempts to break down the human decision process into a series of interdependent and converging phases, to provide statistics and policy implications at each phase, and hence to suggest a quasi-integrative schemata of theme. In so doing, that may be the best that can be achieved: provide a systems model that attempts to maximize a desired goal—namely, the attainment of the greatest number and the best students at reasonable cost into areas of science and technology (S&T). But before going into that let me make a few points with regard to factors affecting career choice (and polices to *influence* that choice).

Over an academic career of about 30 years, I have had upwards of 3,000 students, many of whom I have had the opportunity to know fairly well. The vast majority had, when questioned, little idea why they chose their undergraduate course of study. (In many cases, "chose" was the wrong word, for progression had been quasi-automatic, as much to do with external forces as free will.) They might be able to list a range of factors: "I was good at chemistry at school"; "I liked the teacher" (was this because he was good at the subject, enjoyed praise?); "I didn't want to go into an office"; "I might be able to do some good with my degree"; "The job prospects appear good . . . but I'm not really sure what I want to do . . ."; "My father studies . . ."; "My career master advised me . . ."; "Our school wasn't good at teaching—otherwise . . . ." Or, on the "negative" side: "Science has done so much harm"; "Math is boring"; "The sciences are too hard, too deterministic, little chance for *personal* creativity . . ."; and so on.

But in nearly all cases this might well be construed as post-decision rationalism, cognitive dissonance, or, at best, a compounding of many factors, the dominant one unknown.

Often it was even more entangled than single-factor causation. Thus, S&T was seen to be entangled with industry, and the image of industry might be seen as dirty, inhospitable. If it was career rather than an intellectual discipline that shaped the psychological image (and hence choice), then choice was determined by some forward imagined event rather than rationalized past experience. In such a context, many of the more simplistic standard survey questions regarding career choice become somewhat suspect.[1] At another level, individuals chose a *subject* but delayed choice regarding career. Indeed, in many cases the connection was hardly made! Career was determined by contemporary factors (at the point of, or close to, graduation) such as economic conditions, relative proportion of job openings, and perceived career prospects. Of course by *then* many avenues were closed—it was unlikely a physicist would become an accountant, or a sociology student a research chemist. (Though we must remember that while the former is possible, the latter is not.) Transfers do, of course, occur (my own life experience testifies: a first degree in chemistry, a postgraduate degree in ergonomics and cybernetics, another in physical chemistry/solid state physics, a doctorate in cognitive psychology), but it was until recently comparatively rare.

Perhaps the best that we can hope for in optimizing and encouraging entry into the S&T arena, as this paper seeks to illustrate, is that at *each stage* of the formal educational process we aim, at a macro policy level, to:

- provide the best infrastructure and facilities for instruction and learning environment;

- attempt at the early stages of the educational process to overcome undue cognitive specialization and reject syllabus, certification, and examination procedures that separate individuals into arts and science before they reach FE and HE stages;

- maximize, in whatever ways possible, the participation rate in FE and HE, and encourage flexibility of entry and re-entry;

- provide student financial support wherever possible and reasonable;

- encourage by numerous means [both formal and informal—media, science center (see OSC below) industrial-academic linkage, science-society syllabus inputs] maximum and stimulatory information regarding the positive contributions that S&T makes in all walks of life to encourage the view that S&T can be a warm subject, motivating and personally involving; and

- encourage industrial/academic crossover and wider societal input into syllabus and organizational structure, staff exchange, and challenging field projects at all levels of the educative process, again with an enabling and motivating purpose in mind.

In short, the aim is to maximize fluidity and flexibility in an organizational sense, while fostering latent ability into shaped performance and interest. By following these two broad principles and translating them into a host of human resource policies, we may:

1. maximize the number of entrants into S&T commensurate with societal needs; and

2. encourage the flowering of the best talent (both cognitively and motivationally).

To say that is *not* to emphasize so-called manpower planning or human resource engineering, but to seek a socioeconomic and institutional pathway that is more open, more natural than is presently the case.

A small minority of individuals *know* what they want to do early in life; an even smaller minority are so gifted that there seems no point in doing anything other than following their particular intellectual bent. Even in these cases, support mechanisms can help (identification of gifted mathematicians?—such programs existed in East Europe). However, for the vast majority of a nation's student population, it is a case of introducing policies that minimize dropout and discard, of positive encouragement needs. We might make reference here to Kurt Lewin's "field theory": viz. shaping the external environmental circumstances that contain or influence the individuals' decision threshold . . . an aiding of the mechanisms: economic, organizational, institutional, informational, encouragement, human interest programs, and policies that specifically aim at the removal of the worst obstacles that hinder interest, skills, involvement, and creativity: viz. poor teaching methods, inadequate resources, false information, too early specialization, counterproductive false images regarding S&T . . . .

On this latter point—the image of S&T—there is considerable room for policy maneuver; it may be one of the most important areas in which we can act. The existence and success of the Ontario Science Center (OSC) in Canada[2]—now a world famous Institute—is instructive. In 1967, the Centennial Year of Canada, each province celebrated with a major commemorative project. Ontario's project was the OSC. One of its major *raison d'être* was to attenuate the wrong image of S&T so prevalent among young people in the late 1960s. For many, science had the wrong image: it was environmentally damaging, industry was not nice, and scientific study was not fun. Many students essentially forfeited S&T careers in that at the *critical* decision stage of entry into HE, they based their decision on false or limited information. The OSC sought to short-circuit that problem on a provincial scale (science center, science museum, hands-on participation, university-school linkage, information lectures, or whatever). In my responsibility for the basic science section and the encouragement of industrial involvement, I had a little part to play. In essence, young students previously had little real information about the work of a scientist or technologist on which to base selection of HE subjects (and possibly a career). If they didn't choose the subject, the career was forfeited.

The important policy point here—explored more fully later in this paper—is to provide information and sustenance, image and function *prior to* a critical decision step or a choice point. This should not be ad hoc or piecemeal but part of a well thought out consistent policy. It applies at every phase of the long sequence of events that parallel initial learning, further study, and career choice. There is a progressive ladder and many fall by the wayside at each step who with encouragement or intervention might otherwise not. It is not sufficient to rely solely upon market forces, the hidden hand of the marketplace, and self-optimization theory.

## INTRODUCTION

The case does not have to be made regarding the enormous need in all nations for a highly skilled scientific and technological workforce (or indeed a scientifically literate society in almost all branches of modern life). Numerous governmental and industrial surveys testify to that need, whether in the pure or applied sciences, in engineering, technology, manufacturing, productive or more advanced research areas. Every sector—chemistry, physics, biological sciences, computers and systems areas, information technology, electronics, biotechnology, materials science, all branches of engineering—signals that increasing need.

More problematically there may well be, over a period of time, new and variable demands for the overall portfolio required as new areas (biotechnology, information technology, materials science, advanced systems analysis needs, or whatever) emerge. Local shortages, oversupply in certain areas, and inadequacies in relation to specialized multidisciplinary or interdisciplinary skills may then emerge. (Related to this, there may be delivery or scheduling problems from academe due to temporal-reorientation and infrastructural adjustment requirements—a much neglected area in national policy terms.)

In relation to the above, detailed policy analysis, manpower forecasting, and related analytic studies are continually under way in most nations or major trading blocs. Need is identified; policies of fulfillment are encouraged or enacted. But there are difficulties.

Many studies testify to the difficulties of such human resource or manpower planning. There are, in free market economies, numerous obvious difficulties. For example, individuals are free to choose their own area of study and interest, large time-lags in market adjustment can ensue before market need is satisfied, and the overall production time of technical personnel (from school through higher education) may not marry well with more immediate societal needs. Individuals may reject (for numerous reasons) scientific careers. Schools may be inadequately equipped (in either personnel, equipment, or curricula) to initiate the process of learning (or motivation to learn) at an adequate pace or on an adequate scale. Individuals may observe greater economic (or social) reward in other areas of commercial activity. In short, the adequate delivery at a sufficient level, of appropriate quality, is circumscribed by the wide range of difficulties in a free market economy.

It is not the purpose of this paper to review, categorize, or detail the scale of, form of, or need of S&T human resources; the prime purpose is to examine the reasons why individuals *chose* scientific or technological careers, the ways in which our very limited understanding of that process has or can be translated into policy-suggestive mechanisms, and the extent to which further analysis in this area is required. Necessarily this also behooves us to consider the reasons why individuals either actively reject such careers or encounter barriers and obstacles, and are inadequately supported in the furtherance of following S&T careers; or whether the scale of academic infrastructural support is less than adequate. In commenting upon the latter, we may then be able to seek new pathways of encouragement, new policy-support mechanisms, and new approaches, as well as amplification of existing but insufficiently resourced mechanisms.

Some analysts might argue that the prime lever in ensuring an adequate supply is an economic one: all may be seen in terms of supply and demand curves, relative rates of pay and conditions of work, job security, and socioeconomic status. Such a perspective is woefully inadequate, however. It takes little credence of the need to build up whole delivery infrastructures in academe (often requiring decades of sustenance). It fails to recognize the importance of *selective interventionism*; the encouragement of *particular* skills; the *coordination* of primary, secondary, and tertiary educational systems; and the *nurturance* and selective support of *specialized* research areas. It is not good enough to say that much of this will be taken care of by the private sector; that academe merely has to produce generalists (at all levels); and that fine tuning, extra training, and skill-substitution will then become the prerogative of the marketplace. A biochemist cannot readily become a computer engineer, and a technician is not the same as a highly specialized research chemist.

Thus, much directive planning, in terms of human resources, is required. But, as we have already noted, such planning, such centralized policy and interventionism, is always at the mercy of the vagaries of individual choice and freedom. It is therefore imperative that we understand as fully as possible the factors influencing that choice in order to be able to encourage, respond, attenuate. This, as we shall see, is no easy thing to do, for the long decision chain that leads to and influences career choice permits a large number of variables that need to be controlled or influenced—many of which are not easily amenable to policy or interventionism. (Thus, if *parental* influence, internal psychological satisfaction, or peer respectability play a part, then although policies may be explored to influence these domains, it is no easy task.)

If all of these difficulties did not exist, then most nations would not have experienced the shortfalls and scientific manpower dilemmas that they have in the past—experience therefore pays witness to the difficulties. Having said this, the scale and form of shortfall (if we ignore demographic factors for a moment) varies quite a lot from nation to nation. Germany, Japan, and several southeastern Asian countries encounter fewer difficulties than, say, the United Kingdom. Or, yet again, certain countries (Ireland, the Netherlands, and Italy) have at various times been able to significantly improve their scientific manpower requirements. Also, several LDCs and NICs have achieved oversupply (in part due to the less than fully developed state of their respective formal economies—see especially India). It is not our purpose here to examine such intercultural differences, but there is much potential in such an analysis.

Against the above scale of difficulties regarding adequate manpower provision of S&T skills, perhaps a case could be made that the major task of any society is to encourage what we might call a maximization threshold insurance policy. By this it is meant that at every stage of the educational and wider economic process, sufficient support (teachers, good curricula, number of student places, scholarships, grants, rates of pay, etc.) be made available to satisfy threshold requirements. Once commensurate with individual taste, scientific participation will be maximized. For very rich societies this may be possible; however, it is obviously an expensive route. Nevertheless, it may be the best. Alternatively, we may seek selective positive discrimination: the application of policy levers at what are considered to be important, indeed critical, decision points in career choice, intellectual development (which underpins and precedes career choice), and continuing study. In order to do the latter, we need to consider the factors influencing choice, the shaping and development of scientific personnel in a systematic way. It is to that process that we now turn. As will be seen, it is useful to consider a rather long chain of events. As we identify and to some extent isolate critical stages, we may note (later) policy issues and possibilities for action.

In the sections that follow, the discussion is broken into three main categories or levels. *First*, we provide a sequential step-wise model of the main stages and associated phases, or decision stages, that would seem to influence choice of S&T as a career. At each of the levels there is room for policy enaction. *Second*, we consider some of the difficulties (and supportive literature or research) that focus on each level. This sets the scene for the *third*, the types of policies that are presently being used or explored. In conclusion, we briefly indicate the forms of further research, evaluation, and analysis that need to be undertaken.

## UNDERSTANDING THE MECHANISM OF CHOICE

How can we best capture and comprehend the factors that influence an individual's career choice, in this case scientific or technological? It is patently obvious that there are numerous considerations: school experience, quality and ethos of early instruction in formative years, innate and shaped ability, stimulating experience, external motivational factors, socioeconomic and demographic background, facilities for further education, ease of entry into higher education, market opportunities (both real and perceived), locality, state-incentive, influence of the corporate process in academic life, relative economic reward, ease of transfer from discipline to discipline, educational selection process, retraining opportunity, motivational challenge, peer respect, special support mechanisms that encourage entry, level and form of societal demand (viz., the socio-technical state of development of a society), facilities of compensatory learning (distance learning, modular degrees, external degrees), influence of professional societies, influence of media, and societal, educational, and industrial propaganda, for example. Is it possible to organize such a list into a satisfactory, robust model that leads to a less than random policy apparatus?

We suggest, as illustrated in Figure 1, that it is useful to consider a *sequential chain of events*, critical stages if you will, that to some extent leads to a clearer picture. However, we should not be over-deterministic in our interpretation with regard to sequential dependencies. Nevertheless, there are some obvious points to be made regarding dependence. Thus, if an inadequate level of instruction in science, curricula deficiencies, insufficiency of resource-support, or grossly poor tuition occurs at an early stage, it is less likely that a student will progress into further or higher education in that area. Similarly, if there is little latent ability, it is equally unlikely that an individual will follow S&T as a career. On the other hand, if, in concert with our early remarks regarding maximization threshold, early instruction is good, then latent ability has to be viewed in *relative* terms. Further along the career-choice chain, if rates of pay, societal esteem, career progression possibilities, or the general feel of a particular occupational setting are viewed in negative terms, then despite positive early educational experience, S&T may not be the career choice. Sequential dependencies do therefore exist, to some extent. Indeed, we may, with some justification, be even more mechanistic and note a hierarchical dependency. Thus, the more the conditions of the early stages are improved, the greater the number of

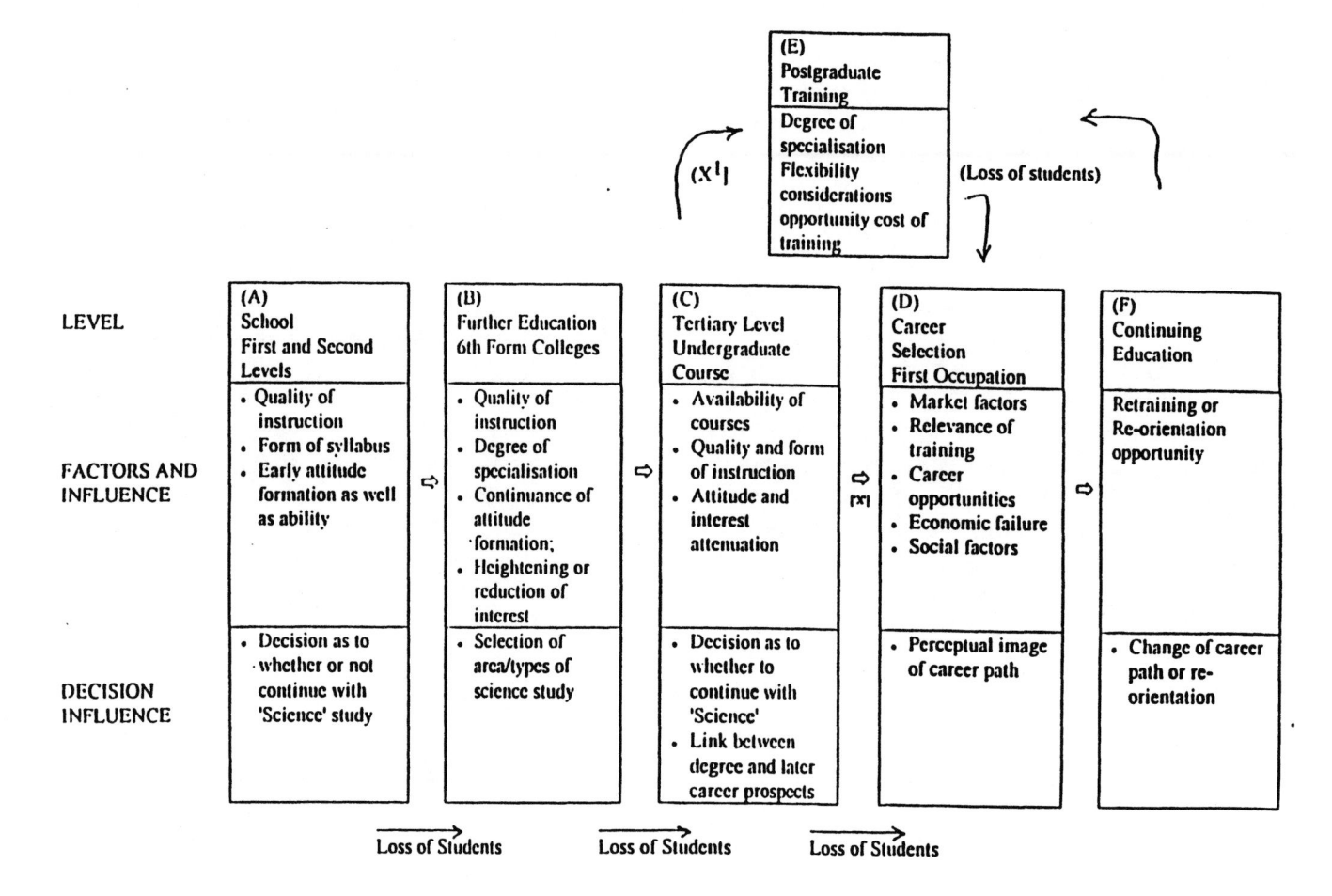

**FIGURE 1** Sequential stages in career selection. (See text for policy appliation at each stage.)

individuals who (may) proceed to the next stage. Obviously some form of selection and filtration is necessary: this implies the value of evaluation, monitoring, examination, standards—all of which will also influence both individual (*internal*) choice and societal (*external*) choice. The policy skill is to get the balance right!

As will be noted at several of the stages in Figure 1, there is a continuing influence upon general attitude to subject matter, the opportunity of heightened or induced interest, and, as we progress from stage to stage, the possibility of a significant reduction in the total manpower resource available.

Stage A—the early school experience—is often critical as a determinant regarding the proportion of males or females who will carry on with scientific studies. Losses at this level almost automatically remove them from the pool available, for later decision, regarding the possibility of a later career in most areas of S&T. (There are considerable limitations at stage F—continuing education—as to the possibility of re-entry or reorientation of career opportunity.) For those who retain a high degree of interest in science at stage B and also have their ability shaped and optimized to a level commensurate with the entry requirements at university level (stage C), there emerges a range of more subtle decision requirements.

Thus, a student may in choosing his or her area of study have a career path in mind that *influences the degree choice*. This can lead to a locking in—an important stage in the career selection process. Alternatively, and more likely (from several surveys), the student may select a degree topic for intrinsic reasons and not be committed at that stage to a career choice trajectory or future path. In this case, the critical decision will be at point X (or possibly X'). Factors, which we shall discuss later, may then be numerous and complex in how they influence the decision process. They may be of a broad socioeconomic nature that influences the general span of occupations available, the degree to which the student has enjoyed or been stimulated by his university instruction and experience, the specificity of linkage between the study period to date and subsequent occupational opportunities, the *perceived* images of career progression possibilities, the extent to which latent ability (in science) has flowered or withered, economic considerations, supply-demand considerations, etc.

In many ways the extra path (see X' in Figure 1) available to postgraduate students introduces an even more complex decision appraisal problem. On the one hand, specialized postgraduate study more closely links the student (if he or she has chosen an S&T area) with the possibility of an S&T career. On the other hand, an extensive period of specialized study sometimes alienates the student from subsequent follow-through. This is often compounded by the relatively high dropout rate in postgraduate research studies. In addition, postgraduate specialization may enhance the likelihood of following an academic career, but in many countries contemporary cutbacks and restructuring of research opportunities may not yield a sufficient range of career opportunities. Meanwhile, the student may have bypassed interest in an industrial career, thereby leading to a decision impasse.

As we shall see below, each of these levels of activity have been subject to study, analysis, and commentary. Equally, each level is (and has been) subject to a wide range of policy-supportive mechanisms aimed at both improving scientific and technological excellence and hopefully influencing in a positive way the decision to follow an S&T career.

However, we must also recognize that to talk of the overall population of students or individuals who might equally follow an S&T career is inappropriate. The total population is not homogenous—different cohorts experience different difficulties. They are subject to different social, economic, academic, and intellectual pressures (e.g., the likelihood of men or women following an engineering or scientific career, or different socioeconomic groups, or different ability groups . . .). This heterogeneity undercuts the value of the simplistic sequential decision procedure indicated above and leads to the need of targeted and specialized policies. (Information technology, biotechnology, and manufacturing managerial skills call for their own specialized support mechanisms.) Policy has to be viewed in that more sub-categorized format. At lower levels of the educational process there are most probably generic, fundamental policies that can be universally applied. However, as we move along the decision ladder, more specialized considerations are required.

What evidence as to the nature of the decision process regarding each level is available? In the next section we consider some of the evidence and wider contributing factors. This is followed by a section detailing contemporary ameliorative policies that, in part, relate to such difficulties and concerns.

## Factors Thought to Influence a Young Person's Decision to Select a Career in S&T: Some Survey and Analytic Data

As we have tried to indicate so far, there are a wide range of factors that are believed to influence the decision to take up a career in S&T.[3] These may include motivational and special interest aspects; economic considerations; image of industry; personality; ability in relation to the scientific and technical field; perception of career opportunities; conditions of work and career prospects; general state of the economy; encouragement by, and influences of, peers, friends, mentors, teachers, and college professors through discussion and advice (a large literature attests to this); conscious and subconscious motivation; attitude to, and image of, S&T itself; availability of *educational* opportunity (which is then a bridge toward later career choice); family background and sociocultural setting; and so on.

We do not intend in this paper to exhaustively review all of these factors—the attached reference list goes some way in signaling the wide range of literature available. But we do provide a selective overview.

An early and exhaustive analysis of *economic* considerations in both a theoretical and practical sense was provided by R.B. Freeman in *The Market for College-Trained Manpower: A Study in the Economics of Career Choice* (1971). A related early study focusing on undergraduate careers in the United States is provided in the exhaustive NORC study *Undergraduate Career Decisions* (J.A. Davis, 1965). Such early studies provide a useful overview of the economic and socioeconomic factors that still pertain to the present time.

We should recognize, however, that the wider socioeconomic circumstances, the wider environmental framework within which an individual makes his or her decision or career choice changes from decade to decade. This goes beyond trade or business cycles. One needs to understand in much more subtle terms than mere supply-demand curves how science, industry, commerce, research opportunity, and *types* of scientific and technological activity *change* in society and status over time. Within this changing context, industry, for example, may seem unattractive at one time, more interesting and attractive at another. Similarly, the actual and perceived opportunity for academic S&T careers fluctuates over the years—as do relative rates of pay and career progression opportunities. Thus, there is not a social constancy in relation to the decision or choice procedure. Times change and, hence, so do the factors.

This dynamic sociological factor is captured well in A. Astin's study *The Changing American College Student: Implications for Educational Policy and Practice* (1991). Unlike several of the more micro, small sample surveys to which we will refer below, Astin's study is based upon survey data that typically involve 250,000 students covering 550 higher education institutions over a period between the late 1960s and the mid-1980s. Astin summarizes the major findings from 24 surveys under 2 headings—career and study plans, and personal values (life goals)—and how these have been changing over time.

He concludes that "between the late 1960s and the mid-1980s American college students became much more focused on material goals and less concerned with altruism and social problems, and these value changes were accompanied by dramatically increased student interest in business careers . . . 'being very well off financially' . . . (plus a reduced endorsement) of 'developing a meaningful philosophy of life.'" These changes are illustrated in Figures 2 and 3.

However, Astin also notes that "during the past two or three years most of these trends seem to have ended or, in certain cases, show signs of reversing direction . . . protecting the environment appears to be the single greatest concern among American college students at the turn of the decade."

The above survey information can carry considerable implications for career choice (in S&T). Many surveys indicate that, for example, industry is not seen as an attractive career, but the shifts in value-change that underpin the Astin data can severely modify our understanding. This is, of course, reinforced by economic recessions, concern regarding job *security*, etc. Thus, the Astin data and more micro-survey data can provided important insight into decision procedure. Similarly, with respect to the environmental concern referred to above, this provides an important policy lever point in S&T careers. If financial reward is becoming an ever increasing choice factor (however other survey data queries this), then S&T careers have to be viewed in that light viz. *relative* rates of pay. However, in policy terms, this may be further complicated by the attraction of economically poorer socioeconomic social cohorts.

Undoubtedly there is a need to recognize that it is not only economic factors that predominate in the choice of S&T career. Fuller (1991) examines the problem of scientific skills shortages in her study *There's More to Science and Skills Shortages than Demography and Economics: Attitudes to Science and Technology Degrees and Careers.*

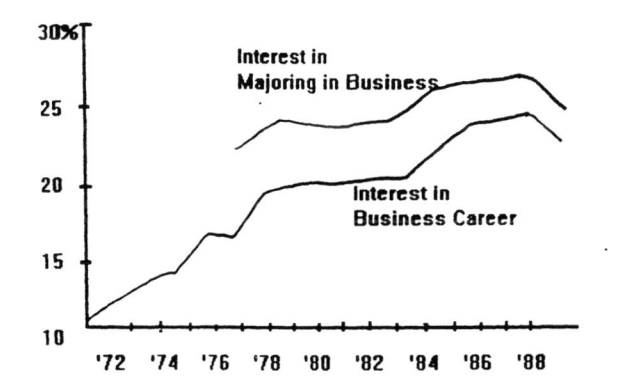

SOURCE: Astin, 1991.

**FIGURE 2** Freshman interested in business.

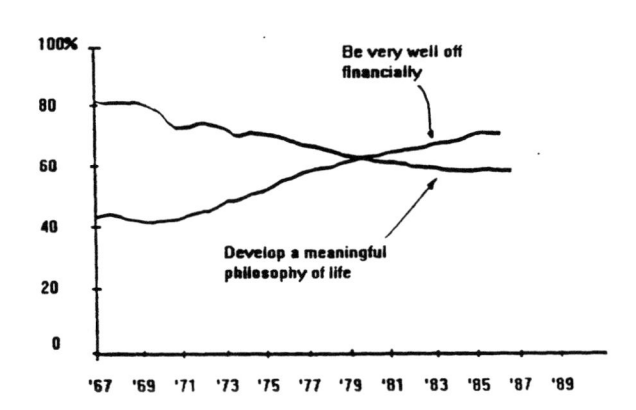

SOURCE: Astin, 1991.

**FIGURE 3** Contrasting changes in two values (rated as "Essential" or "Very Important").

Based upon detailed interviews, the paper "explores first-year university and polytechnic students' attitudes to degrees and careers in S&T and the factors influencing their choices to pursue S&T or turn to alternative non-S&T areas. Students' choices and decisionmaking patterns provide pointers for those in schools, higher education, industry, and government who wish to understand why some students are deterred from studying S&T and who may be interested in making changes that could encourage underrepresented groups to pursue S&T further."

Fuller notes "skills shortages are one of the most talked about difficulties that British industry currently faces. Those concerned expect the challenge to increase as the demographic decline and the European Single Market begin to bite, and rapid technological change and intense global competition persist." And "adding to their concern are statistics (indicating) that only about a quarter of science graduates graduating in 1987 entered scientific employment. Furthermore, the total number of S&T students graduating in the United Kingdom has compared unfavorably in recent years with the numbers produced by West Germany, the United States, and Japan" (Prais, 1989).

In relation to these difficulties Fuller examined, through interviews, five areas:

1. The nature of the decisionmaking process

2. Vocational awareness and vocational flexibility

3. Why students with S&T "A" levels turn to non-S&T degrees

4. The effects of perceived ability on choices

5. Factors affecting female students' choices

### The Nature of the Decisionmaking Process

Here "the data indicated that two major influences, perceived academic strength and enjoyment, had affected both S&T and non-S&T students' academic choices. The students were asked to reflect carefully on other factors, such as school type, career guidance, teachers, parents, and friends, but, in almost all cases, participants insisted that success and interest formed their principal decisionmaking rationale." Early learning experience and academic attainment clearly played a big part. (This reinforces a central argument that I am emphasizing here that early school experience is an important factor in later career choice.)

**Vocational Awareness and Vocational Flexibility**

Fuller posits, "In thinking about S&T skill shortages in industry it was interesting to discover at what point student's awareness of vocational options started to impinge on choices." Responses indicated that at 14-16 years old, decisions were not career-driven. At 18 it was a greater factor, but enjoyment and academic success were still the dominating factors—not an assessment of "opportunity structures within the labor market."

But students tried to keep options open (vocational flexibility), for example, by opting for general (not specialized) chemistry courses—which were seen to have "high exchange value in the labor market." Equally broad-based S&T topics would allow entry into *non*-S&T areas. The main point is that in terms of personal development, the ability to respond to a variety of career opportunities was retained. (A policy implication here is the importance of flexible postgraduate top-up or transfer of courses for those who may later wish to return to a S&T career.)

**Why Students with S&T "A" Levels Turn to Non-S&T Degrees**

Two categories were observed: those who consider that S&T degrees and careers would limit their personal development and those who wanted degrees but whose interest in S&T had waned. The first group often perceived themselves as intellectual and academic high flyers. They saw the S&T degree timetable as too full and inflexible, not allowing personal autonomy. They often saw S&T careers, particularly in industry, as constraining, promotion-limited, not challenging enough, and not permitting either self-employment or sufficient opportunity to articulate ideas and views.

As Fuller notes, the reality is different from employed graduate surveys and contradicts this assessment of career conditions:

> We see here an important policy need in informational linkage between schools, university, and industry.

Regarding the second group (waned interest), Fuller points to the restructuring influence of lack of flexibility in British higher education: more cross-disciplinary options should be provided.

**The Effects of Perceived Ability on Choices**

Science subjects, particularly mathematics and physics, were perceived as more academically demanding where only the most intelligent can succeed. Thus, young, competent, but not brilliant, students were discouraged from following S&T courses in HE. Nevertheless, as Fuller points out, employers want a range of general strengths and good interpersonal skills, and will provide extra training. There is therefore "a gap between what some in education assume students need to pursue S&T and what large employers were looking for in their employees." Again, we see important policy implications here.

**Factors Affecting Female Students' Choices**

Female students often reported they were in a minority in chemistry, physics, and math classrooms; timetabling was seen to be against traditional female study areas.

Very few had been exposed to equal opportunity campaigns such as Women in Science and Engineering (WISE) or taken part in special events aimed to encourage girls into S&T. (Even the timing of these events was too late—after options had been chosen, *again we see the importance of the decision chain* in following a career.) The problems of combining career and family were constantly emphasized as a factor in career choice. Fuller suggests that seven major factors discourage many potential S&T candidates from choosing S&T options and subsequent S&T careers:

1. You have to have a certain sort of mind to achieve in S&T at degree level.

2. S&T subjects, especially mathematics and physics, are uniquely demanding academically.

3. S&T degrees and careers do not require effective communication and interpersonal skills.

4. S&T degrees and careers are insufficiently

stretching and rewarding for academic and intellectual high-flyers.

5. Some S&T degrees lack vocational flexibility.

6. S&T subjects at school and in higher education are male preserves and promote masculine values.

7. It is particularly difficult to combine S&T careers with having a family.

Student responses indicated that these messages influenced individuals within three particular groups—high-flyers, the less-than-brilliant, and girls—to opt for non-S&T degrees.

(All of the above difficulties are subject to ameliorative policies. In addition, surveys of conditions in industry do not tally with student perceptions; this signals the importance of better academic-industrial linkage programs. The Enterprise of Higher Education Initiative aims to provide more students with the opportunity to gain industrial experience through project work, as does the Credit Accumulation Scheme. At the postgraduate level, such programs as the Teaching Company Scheme and the CASE Awards have equal importance at the research level.)

Career prospects have also to be viewed across a wider horizon. Thus, "(engineers) were optimistic that the advent of the Single European Market would bring about change in the UK because S&T workers would have increased opportunities to work in Europe . . . with benefit from comparatively higher status and salaries . . . and scientists point out that international mobility within their profession has been common in recent years." But the HE system has an important role in this. Fuller argues that her study has indicated that "S&T courses discourage suitably qualified students for two principal reasons: first, they are perceived as too demanding and second, as limiting autonomy and therefore personal development." The need is to review the process and content of courses to attract a wider variety of students and introduce more combined studies (e.g., engineering with business studies, encourage project work and academic-industrial exchange programs, and overall synergy and communication).

The above comments depend, in the main, upon attitude surveys; however, preference is also indicated in an empirical sense at the macro level by the numbers entering various fields of higher education and the choice of first destination of career. The first category *influences* career choice (though we have just seen that it does not automatically confirm it), while the second category tells us more of preference commensurate with market conditions and hence sectoral opportunity.

At an EC level, various OECD and EC publications reveal gross statistics. For example, Figure 4, taken from *Europe in Figures* (Eurostat, 1992), indicates the changing pattern from 1970-1972 to 1988-1989.

As can be seen, both engineering and natural sciences have failed significantly in percentage terms over that time, although the absolute number has increased.

However, "the social sciences accounted for about one student in seven in 1988-1989 compared with a figure of one in nine in 1970-1971. Arts students headed the list at the beginning of the 1970s (16.5 percent) but had dropped to second place in 1988-1989 (14.1 percent)." We therefore begin to see a pattern of expressed preferences especially toward the social sciences here.

In terms of first career preference, Bee and Dolton (1990) provide for the UK extensive data. They provide an overall national analysis of the initial career pattern of *all* university graduates over time and across faculties for the period 1961-1962 to 1986-1987, based upon *First Destination Returns,* which has been compiled nationally by the UGC and the USR. The authors identify the main trends and seek to explain these by relating them to changes in the occupational structure of the UK labor force. They conclude, "It is demonstrated that, while there is some correspondence between the two, the relationship is neither simple nor exact . . . the pattern of graduate first destinations has depended not only on structural changes in the economy but also on a range of institutional and market forces that have operated specifically on the demand for, and supply of, highly qualified manpower."

How did S&T fare in this? What does this tell us regarding changing patterns of preference? Figures 5 through 11 are of particular interest.

As Bee and Dolton note, "shifts in graduate career preferences, affecting the relative attractiveness of different occupations, have undoubtedly occurred." Important is the "persistent failure of industry to attract arts and social science graduates, the dominance of sciences in research, and the growing attraction of commerce for scientists and technologists . . . as is the surprisingly high (though falling) percentage of scientists who embark on teacher training."

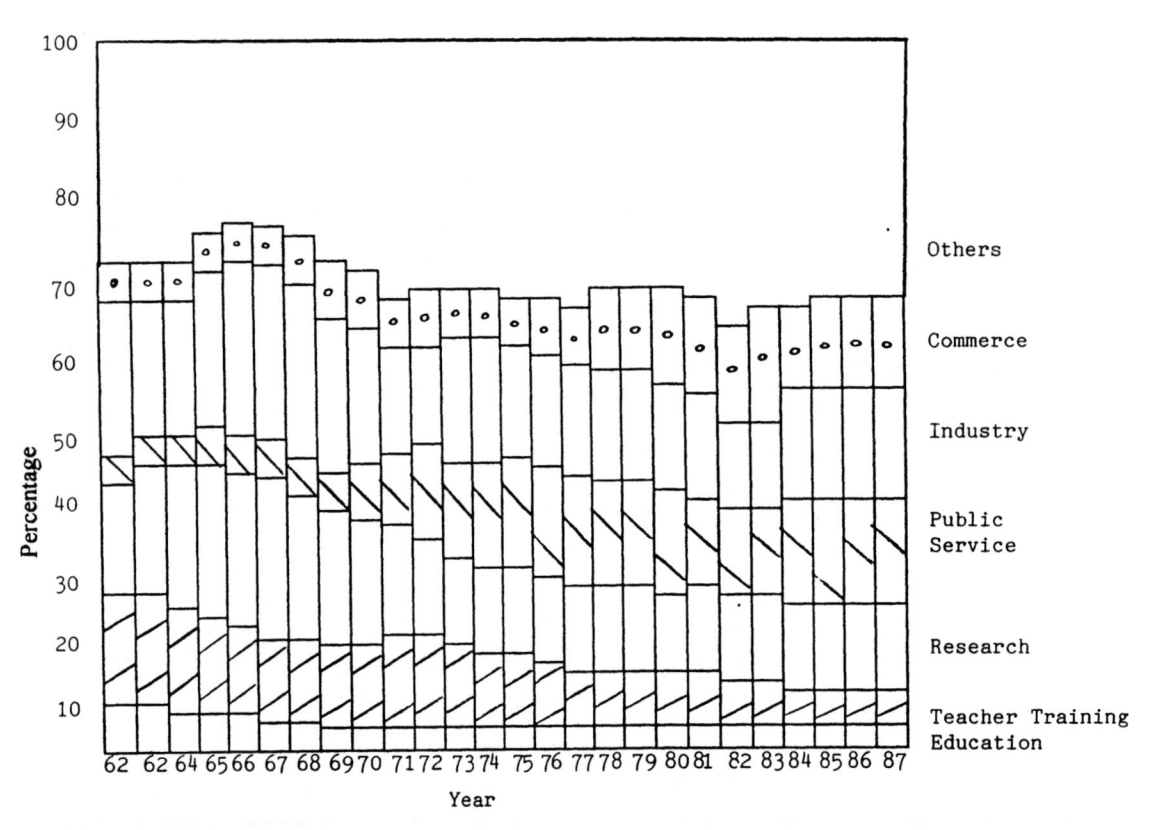

SOURCE: Eurostat, 1992.

**FIGURE 4** Part-time and full-time third-level students by field of study in EC (percentage).

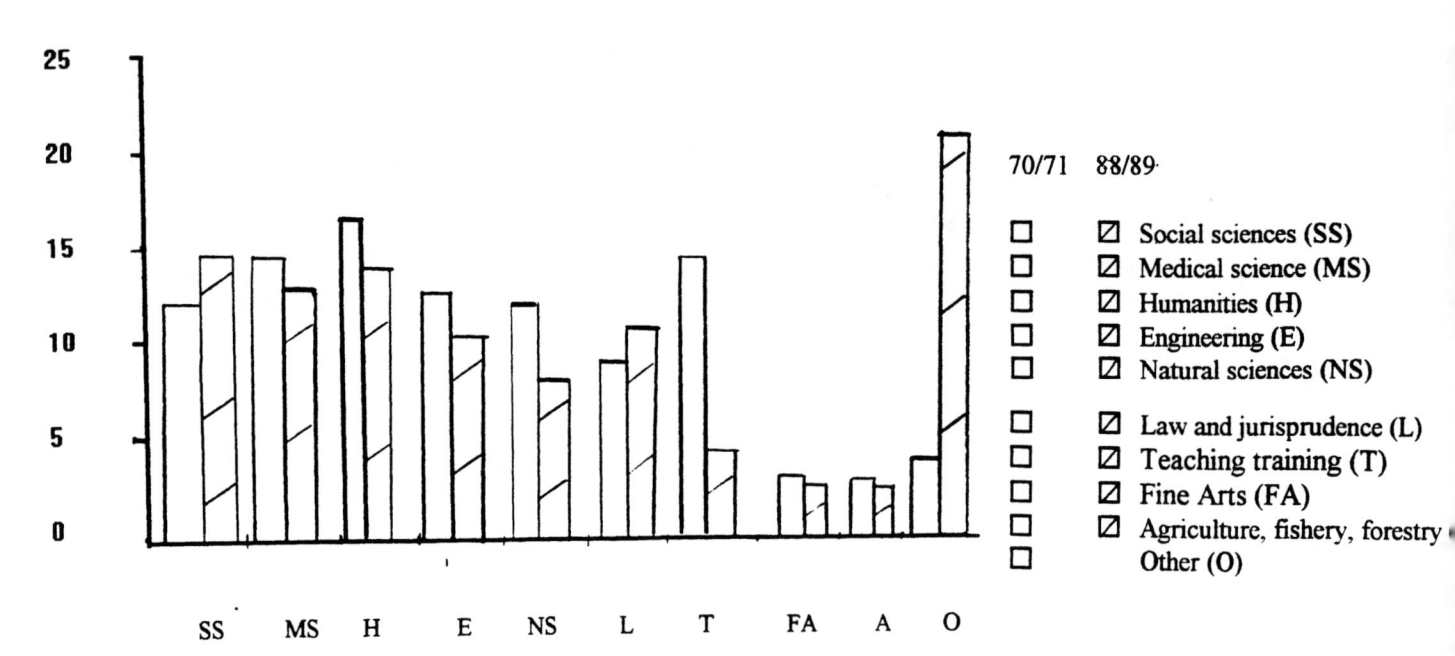

SOURCE: Bee and Dolton, 1990.

**FIGURE 5** Graduate first destinations: cumulative percentage (1962-1987).

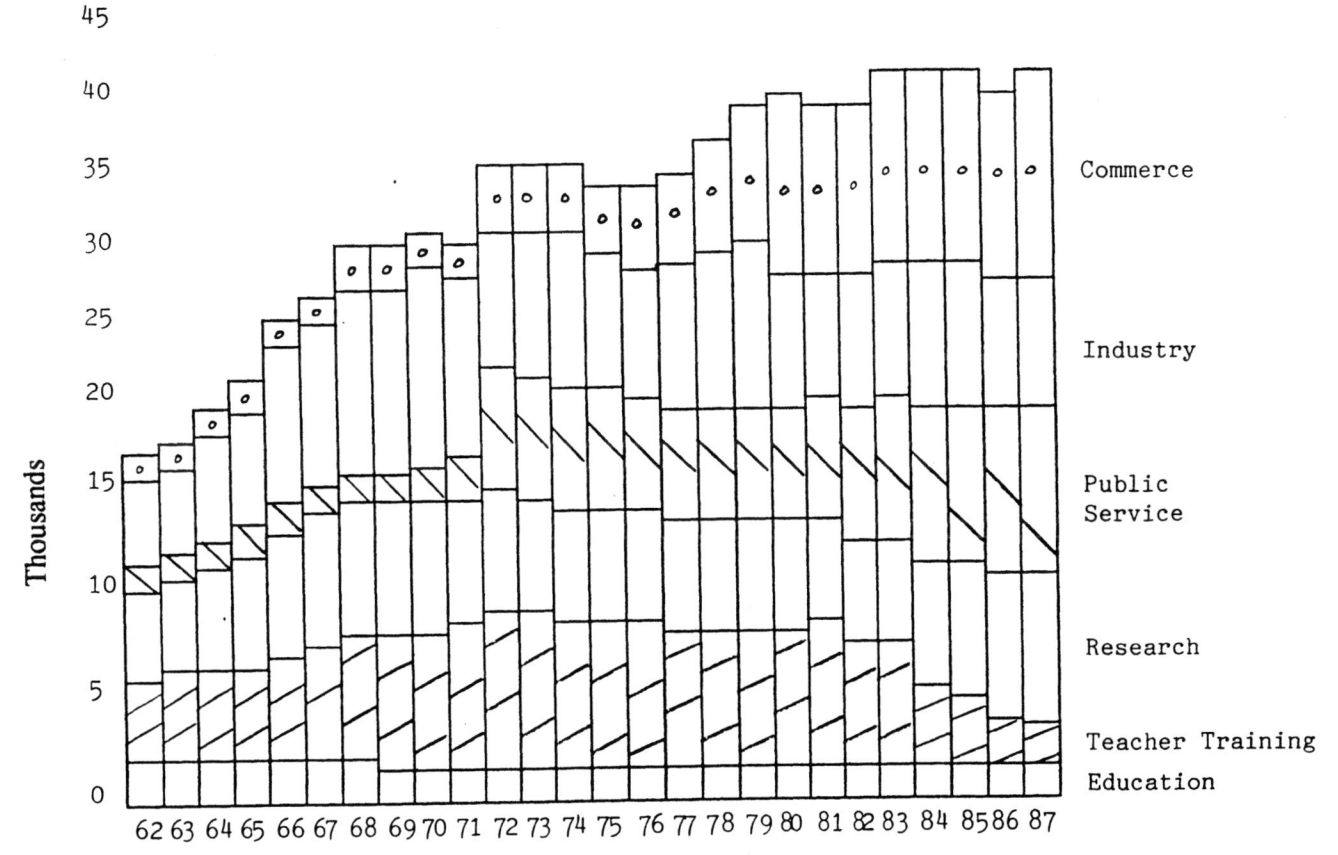

SOURCE: Bee and Dolton, 1990.
**FIGURE 6** Graduate first destinations: numbers (1962-1987).

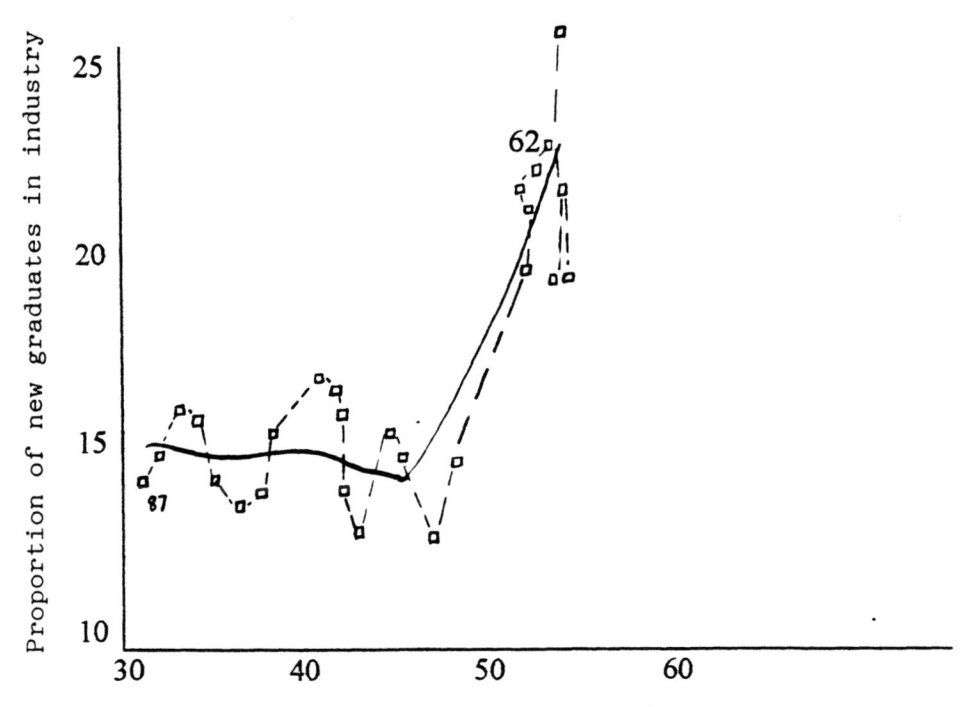

**FIGURE 7** First destination of university graduates: industry.

*CAREERS IN SCIENCE AND TECHNOLOGY: AN INTERNATIONAL PERSPECTIVE*

SOURCE: Bee and Dolton, 1990.
**FIGURE 8** First destinations of university graduates: industry.

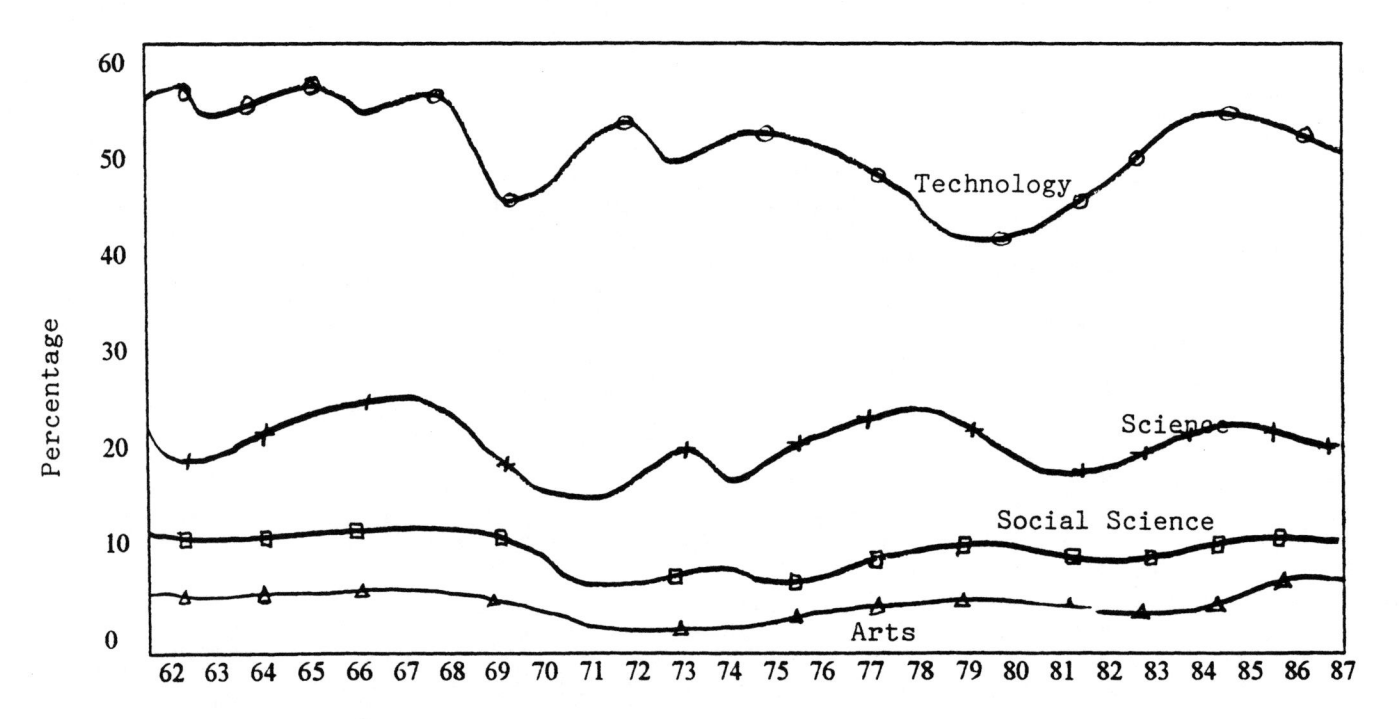

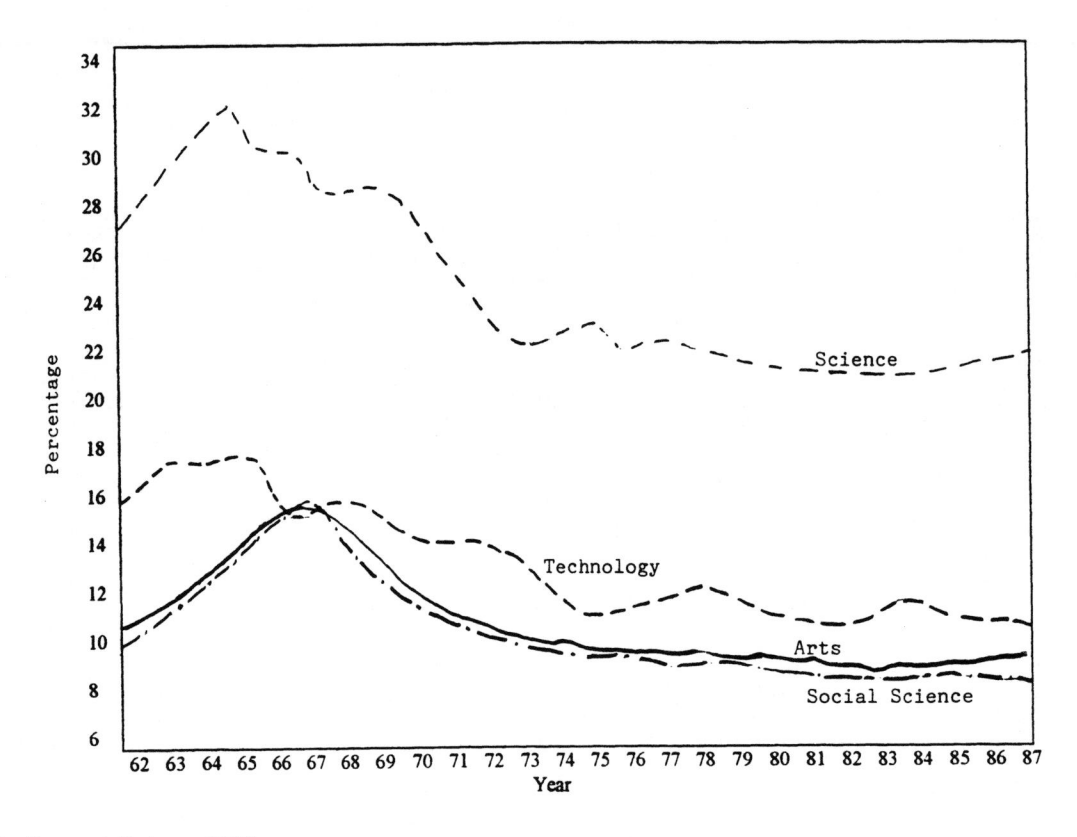

SOURCE: Bee and Dolton, 1990.
**FIGURE 9** Research percentage by faculty (1962-1987).

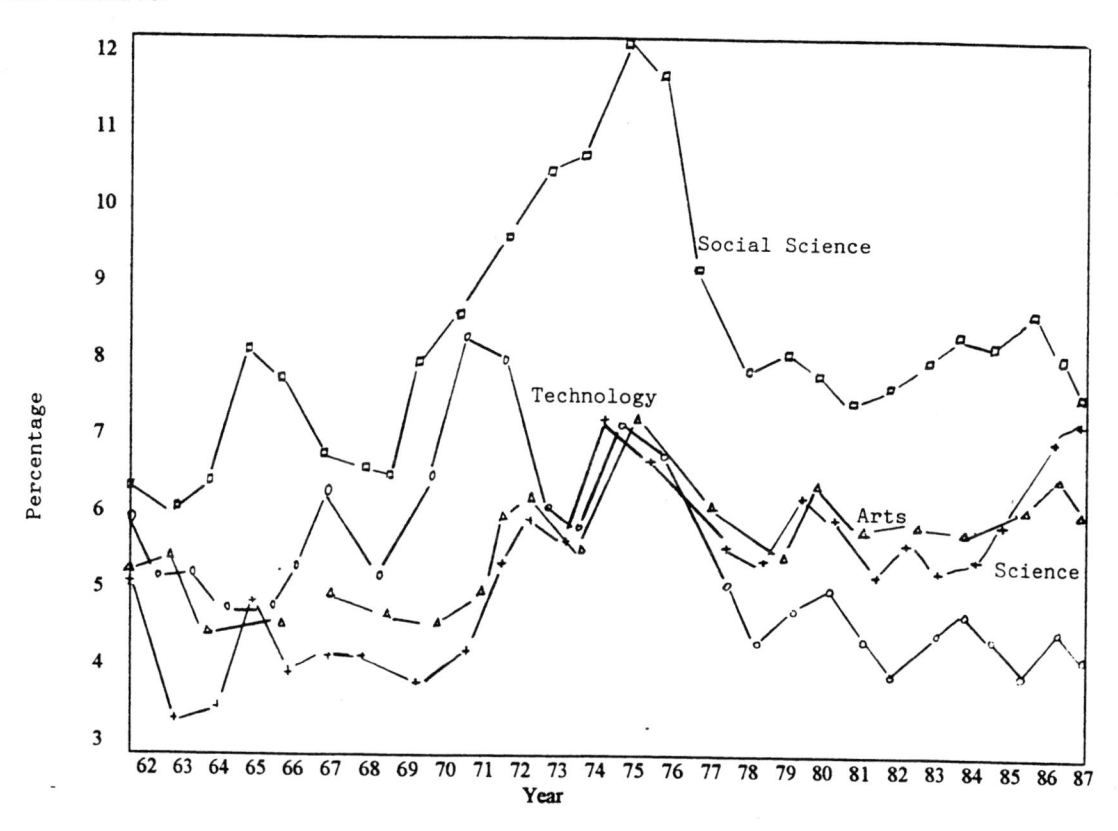

SOURCE: Bee and Dolton, 1990.
**FIGURE 10** Public service: percentage by faculty (1962-1987).

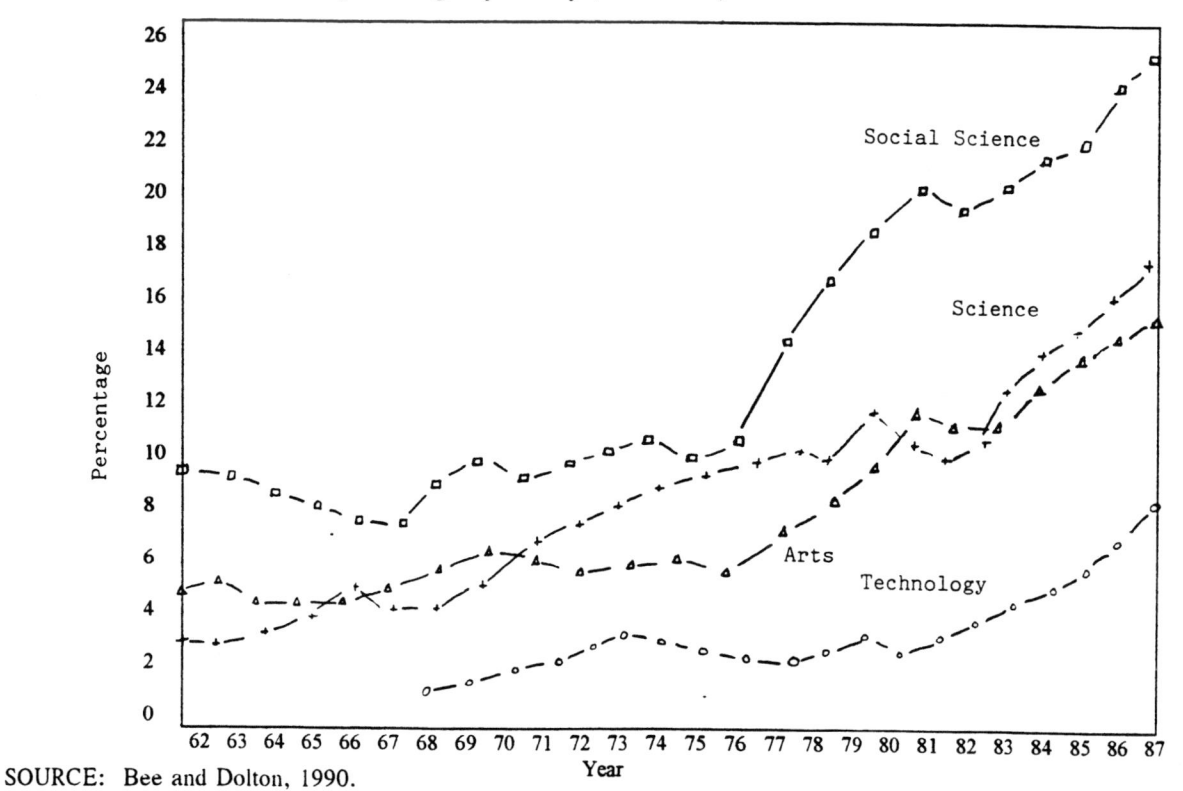

SOURCE: Bee and Dolton, 1990.
**FIGURE 11** Commerce: percentage by faculty (1962-1987).

It is important, however, to recognize that there are considerable international differences that complicate any robust analysis. The UK is not necessarily typical of Europe at large. For example, in recent years the UK government has introduced polices to expand university intake[4] for a variety of reasons. However, under free market conditions, the uptake in various faculty areas has been extremely uneven as shown in Figure 12.

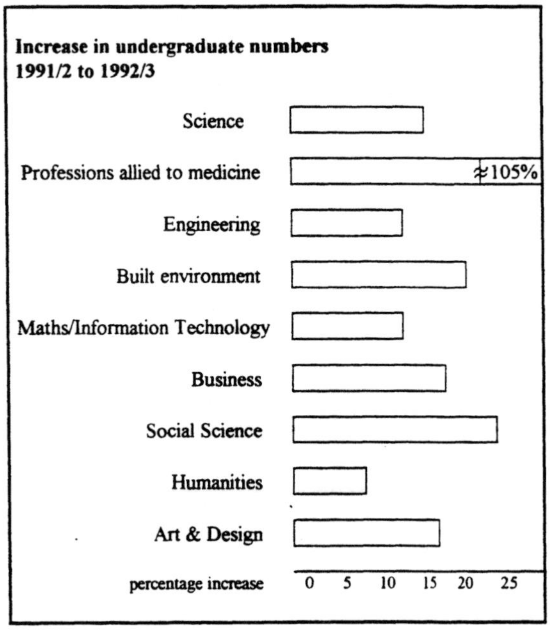

```
Increase in undergraduate numbers
1991/2 to 1992/3
                            ≠105%
            Science          ▭
Professions allied to medicine ▬▬▬▬▬
         Engineering         ▭
      Built environment      ▭▭
Maths/Information Technology ▭
           Business          ▭▭
       Social Science        ▭▭▭
          Humanities         ▭
         Art & Design        ▭▭
    percentage increase   0  5  10  15  20  25
```

SOURCE: THES, 1992.

**FIGURE 12** Increase in undergraduate numbers, 1991/2-1992/3.

[As can be seen, increased intake into engineering and science has been disappointing. Policies are now being considered through the unit of resource (effectively the income universities received in relation to student area of study) to be reduced in classroom-based areas (arts and humanities) as a means to encourage recruitment to engineering and technology courses (THES, p. 3, December 4, 1992).]

But the UK may not be typical of the EC at large as shown in Figures 13 through 15, which reveal different patterns of career preference (and educational provision).

We have indicated, so far, that career aspiration is a function of a wide range of personal, institutional, and socioeconomic factors. Are more precise survey details available? Boys and Kirkland (1988) undertook a survey of 6,000 final year undergraduates in 1982 (of which 58 percent were completed and analyzed). A follow-up survey in 1985 yielded a 59 percent response rate. In their text *Degrees of Success: Career Aspirations and Destinations of College, University, and Polytechnic Graduates*, they examine qualifications and career aspirations, early destinations, career opportunities and prospects, realization of aspirations, level of income factors, and retrospective evaluation of degrees.

In terms of aspiration and career they note, inter alia, the data indicated in Tables 1 and 2.

In terms of ranking various factors in relation to long-term career plans, Table 2 is informative. However, Boys and Kirkland are fairly sanguine regarding its interpretation. Thus, they state:

> As Table 2 shows, there was not consensus among respondents that any of the career aspirations presented to them were very important. In the highest ranking, 59 percent of all final year undergraduates in 1982 felt that a job that offered good long-term opportunities was very important, 31 percent fairly important, and only 10 percent not important to them.

> The majority of graduates attached at least some importance to 17 of the 22 items. Among these, the importance attached to a high starting salary, ranked 17, was much lower than that attached to a high future salary, ranked 2. Only 10 percent considered a starting salary very important, and 44 percent felt it to be of little importance, compared with 40 percent and 18 percent respectively for long-term earnings. Half of all undergraduates attached little importance, and very few a great deal of importance, to careers that offered social prestige and status or good pension plans.

> While there were correlations between particular aspirations and the type of institution attended or subject studies, these were in most cases weak. Largest subject variations were found in the importance attached to working in industry. Here subjects could be classified into three groups. Four-fifths or more of lawyers,

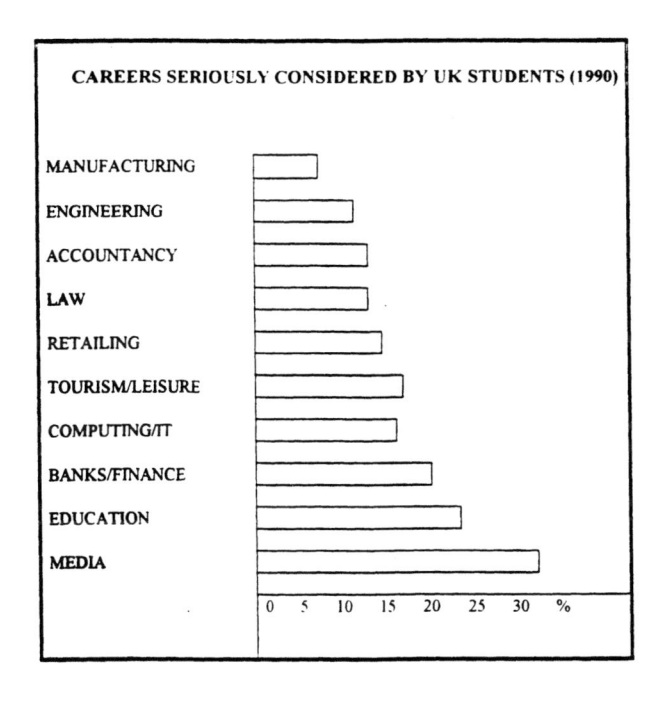

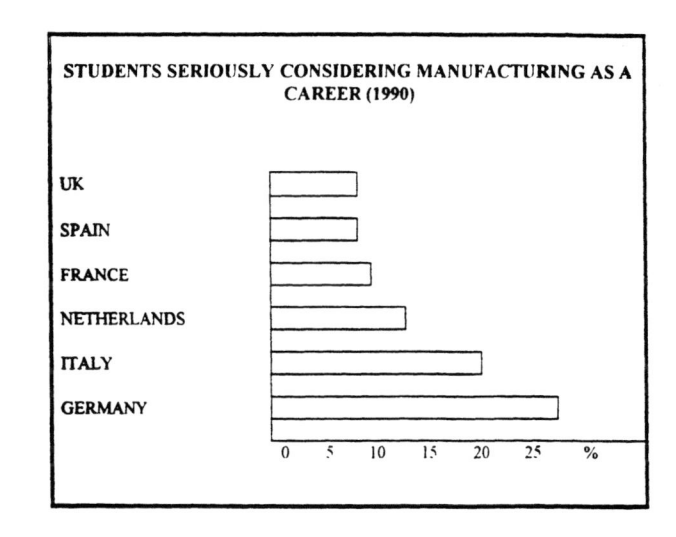

SOURCE: HMSO, 1991.

**FIGURE 13a** Careers seriously considered by UK students (1990).

SOURCE: HMSO, 1991.

**FIGURE 13b** Students seriously considering manufacturing as a career (1990).

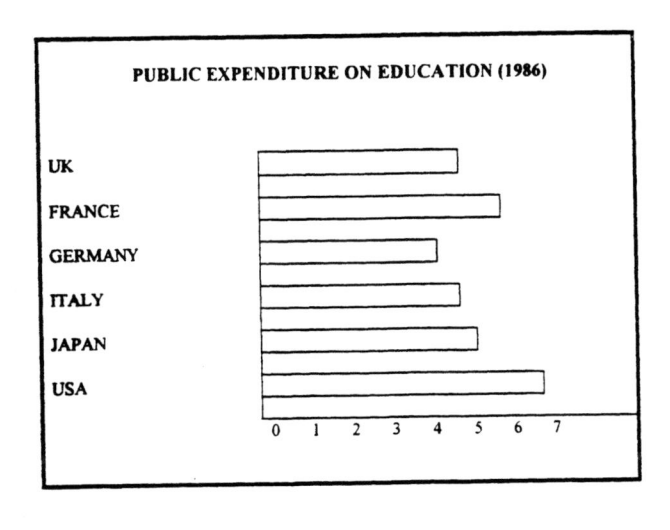

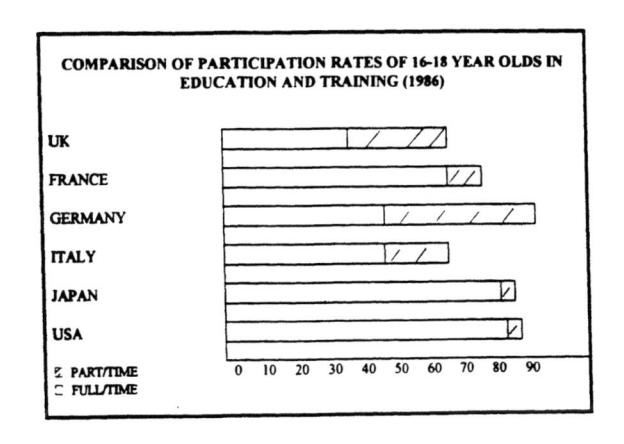

SOURCE: HMSO, 1991.

**FIGURE 14a** Public expenditure on education (1986).

SOURCE: HMSO, 1991.

**FIGURE 14b** Comparisons of participating rates of 16- to 18-year-olds in education and training (1986).

## Degrees Awarded 1986 (as % of age group)

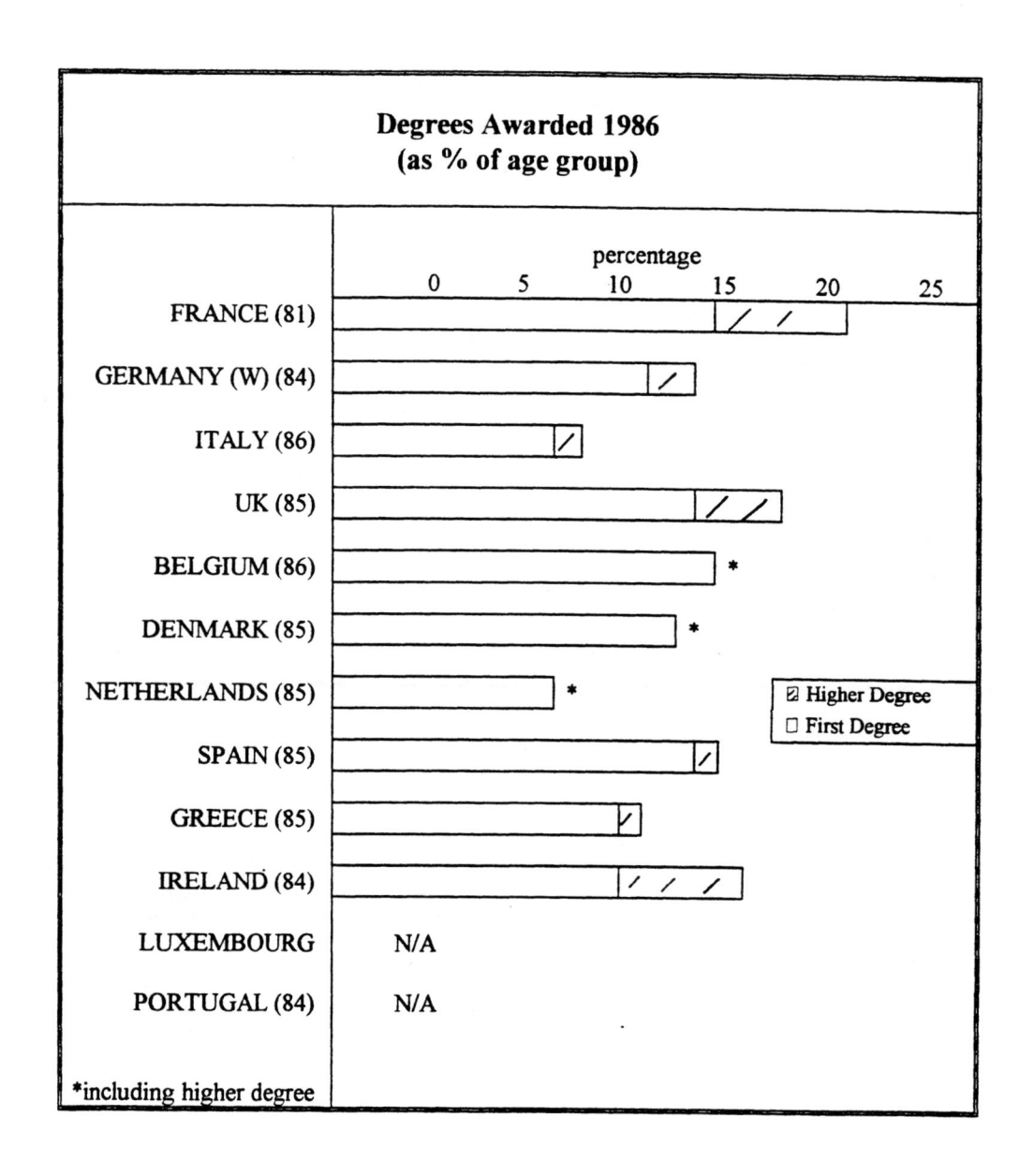

| Country | Percentage |
|---|---|
| FRANCE (81) | ~20 (incl. higher degree portion) |
| GERMANY (W) (84) | ~13 |
| ITALY (86) | ~7 |
| UK (85) | ~17 |
| BELGIUM (86) | ~15 * |
| DENMARK (85) | ~13 * |
| NETHERLANDS (85) | ~7 * |
| SPAIN (85) | ~14 |
| GREECE (85) | ~11 |
| IRELAND (84) | ~15 |
| LUXEMBOURG | N/A |
| PORTUGAL (84) | N/A |

Legend: ▨ Higher Degree   ☐ First Degree

*including higher degree

SOURCE: IMS, 1990.

**FIGURE 15** Degress Awarded, 1986 (as percentage of age group).

**TABLE 1** Career Aspiration as a Function of Discipline

71% of Lawyers wanted to pursue careers in the legal profession
65% of Engineers wanted to pursue careers in the engineering profession
38% of Commercial graduates wanted to pursue careers in the finance profession
55% of Chemists wanted to pursue careers in scientific research, design, or development
40% of Mathematicians or Computer Scientists wanted to pursue careers in the management services (associated with computer science)

SOURCE: THES, 1992.

**TABLE 2** 1982 Career Plans Ranked by Average Score Based on the Question: How important are the following factors in your long-term career plans?

| Factor | Rank | Score |
| --- | --- | --- |
| A job that gives me good long-term career opportunities | 1 | 2,4873 |
| A high future salary | 2 | 2,2236 |
| The opportunity to be creative and original | 3 | 2,1315 |
| The opportunity to use knowledge gained on my degree course | 4 | 2,0969 |
| A job with a lot of responsibility | 5 | 2,0812 |
| A suitable geographical location | 6 | - |
| A job that will give me the opportunity for rapid promotion | 7 | 2,0228 |
| The opportunity to use the skills I acquired on my degree course | 8 | 2,0210 |
| Long-term job security | 9 | 2,0172 |
| Work in which I'm independent of supervision | 10 | 1,9143 |
| A job with flexible working hours | 11 | 1,8594 |
| A job with good fringe benefits | 12 | 1,8075 |
| The opportunity to travel and work overseas | 13 | 1,7949 |
| A job in which I will work as part of a team | 14 | 1,7709 |
| A job that is concerned with helping others | 15 | 1,7681 |
| A career that will allow me to move from job to job | 16 | 1,7532 |
| A high starting salary | 17 | 1,6656 |
| A job with social prestige and status | 18 | 1,6252 |
| A job with a good pension plan | 19 | 1,5978 |
| A job in industry | 20 | 1,5479 |
| Probability of eventual self-employment | 21 | 1,5298 |
| Work that will be mainly out-of-doors | 22 | 1,2296 |

SOURCE: Boys and Kirkland, 1988.

graduates studying other arts and humanities, social sciences and history attached little importance to industry. A little over two-fifths of economists, mathematicians or computer scientists, chemists and other scientists gave it at least some importance, although only between 14 and 18 percent felt that such a career was very important. Finally, over half of the engineers and commercial graduates attached at least some importance to industry. Of these two subjects, engineering students were the more industry motivated: 43 percent felt it very important, and 34 percent fairly important, compared with 24 percent and 30 percent from commercial subjects. As discussed below, these variations were also associated with institutional factors.

Commercial, engineering, and law undergraduates were more likely than others to attach importance to those aspirations that could be described as 'extrinsic rewards' (Niessen and Pescher, 1981). These included financial and other material benefits (a high starting salary, good pension plan, high future salary, and good fringe benefits) and others such as social prestige and status and the opportunity for rapid promotion. History, other arts and humanities and social science undergraduates were less likely to attach importance to such factors.

More specifically, in relation to industrial careers and the need to attract the "brightest and highest calibre products of our Universities and Polytechnics into the wealth-creating segment of Society," the (UK) Committee for Research into Public Attitudes, under the chairmanship of Lord Plowden, undertook a detailed study based upon 1,007 "lengthy personal interviews with undergraduates and polytechnic students in 55 carefully selected and stratified educational institutions" (see *Attracting the Brightest Students in Industry*, 1985). Of particular interest is that the numerous tables of detailed results are categorized and tabulated not only by subject area of study, but also by expected degree class and attitude to industry. We cannot present all the results here, but note several of the most significant findings:

- The importance of job satisfaction/enjoyment . . . "overwhelmingly anticipation of enjoying the job is what points graduates in a particular direction."

- Although, at first, high starting salaries were not that important, "we suspect that, in practical terms, money is a powerful attraction."

- The prospect of good training and job experience facilities is a key motivator.

- A high score goes to good career prospects; they want a job to have a series of opening doors.

- Career advisory staff have a large influence on influencing career choice. (We shall refer to this shortly.)

Some of the findings are reflected in Tables 3-5.

We indicated above the importance of career advisory staff as facilitators for S&T students to undertake S&T careers. Connor et al. (1992) provided a highly detailed review and analysis of this area in the report to the Engineering Training Authority, *A Careers Service for Engineering*. Numerous tables, descriptants, and survey results are provided. An overview is given in Figure 16.

The study emphasizes that it is of considerable importance to maintain such inputs right through the whole educational life cycle. (See Table 6.)

Such a facility then relates to the sequential decision steps described in earlier sections of this paper. Occupational selection should not be (and indeed in many ways is *not*) a random process. It is influenced by a whole chain of events over many years. That is what sequential dependency means. Information provision, continuous good instruction and learning facilities, maximizing student intake at each level, and close societal-academic informational links and exchanges are all part of that cumulative process. Each is potentially subject to many policy inputs.

## Policies to Improve S&T Literacy
## (and Possible Selection of S&T as a Career)

As will be seen in Table 7, a wide range of policies, supporting mechanisms, are presently being used to improve scientific instruction, to increase interest and motivation regarding S&T areas, and, either directly or indirectly, to increase the chances of

**TABLE 3** Question: What are your reasons for thinking that a particular career is what you want?

| Base: All who have decided on careers | All 682 % | Computing 67 % | Teaching 58 % | Non Academic Research 31 % | Unspecified Engineering 55 % | Electronic Engineering 60 % | Civil/Struct Engineering 48 % | Law 40 % | Management 35 % | Accounts Finance 41 % |
|---|---|---|---|---|---|---|---|---|---|---|
| Am interested/would enjoy it | 45 | 57 | 43 | 68 | 44 | 52 | 46 | 50 | 46 | 27 |
| Can use degree subject | 18 | 28 | 12 | 32 | 16 | 25 | 13 | 20 | 23 | 7 |
| Intellectual challenge | 12 | 12 | 5 | 10 | 4 | 12 | 19 | 10 | 14 | 15 |
| Would be good at it | 10 | 7 | 16 | 3 | 13 | 12 | 10 | 15 | 11 | 10 |
| Have had experience | 10 | 18 | 9 | 13 | 13 | 7 | 2 | 8 | 6 | 10 |
| Job satisfaction | 9 | 9 | 10 | 3 | 4 | 8 | 13 | 10 | 6 | 2 |
| Earn a lot/financial reward | 9 | 16 | 5 | 3 | 5 | 8 | 8 | 15 | 9 | 22 |
| Good promotion prospects | 8 | 10 | -- | -- | 5 | 15 | 4 | 10 | 17 | 32 |
| Involves working with people | 7 | -- | 31 | 10 | 5 | -- | 2 | 5 | 11 | 5 |
| Socially worthwhile | 5 | 1 | 7 | 10 | 4 | 2 | 8 | 3 | -- | -- |
| New/expanding area | 5 | 19 | 2 | 6 | 4 | 13 | 2 | -- | 3 | -- |

SOURCE: "Attracting the Brightest Students in Industry" (Opinion Research and Communication, 1985)

**TABLE 4** Question: Here are some of the attributes that might make an organization attractive to graduates. Please put the following aspects in order of choice (percent making each aspect first or second choice).

|  | EXPECTED DEGREE ||| ATTITUDE TO INDUSTRY || ENGINEERING |||| SUBJECT |||||||
|---|---|---|---|---|---|---|---|---|---|---|---|---|---|---|---|
| BASE | ALL (1007) | 1st/2nd (663) | Other (266) | Positive (675) | Neg. Neutral (330) | Gen Mech Prod (94) | Electric Electron (117) | Aero (56) | Other (122) | Math Physics (92) | Biology Chem (109) | Computers (77) | Bus, Econ, Accts (99) | Social Studies (80) | Humanities Law (136) |
| Offering job satisfaction | 69 | 69 | 70 | 68 | 70 | 74 | 72 | 63 | 58 | 69 | 82 | 65 | 65 | 69 | 73 |
| Good training and job experience facilities | 43 | 43 | 44 | 44 | 43 | 46 | 49 | 45 | 51 | 45 | 38 | 45 | 44 | 35 | 39 |
| Offering career prospects | 43 | 43 | 40 | 45 | 37 | 48 | 45 | 36 | 40 | 42 | 46 | 34 | 39 | 38 | 45 |
| Salary and pension arrangements | 16 | 16 | 18 | 17 | 15 | 8 | 16 | 24 | 21 | 22 | 11 | 26 | 15 | 17 | 13 |
| Working for a successful and prestigious organization | 12 | 13 | 12 | 12 | 12 | 14 | 9 | 9 | 10 | 16 | 19 | 10 | 14 | 10 | 14 |
| Offering a varied choice of location | 11 | 11 | 12 | 9 | 16 | 8 | 7 | 6 | 18 | 7 | 7 | 9 | 16 | 18 | 11 |
| Good fringe benefits | 4 | 3 | 5 | 3 | 5 | 1 | 2 | 7 | 1 | 3 | 4 | 6 | 4 | 5 | 4 |

SOURCE: Opinion Research and Communications, 1985.

**TABLE 5** Question: Within job satisfaction, which two individual items do you consider important? (Percentage)

|  | EXPECTED DEGREE ||| ATTITUDE TO INDUSTRY || ENGINEERING |||| SUBJECT |||||||
|---|---|---|---|---|---|---|---|---|---|---|---|---|---|---|---|
| BASE | ALL (1007) | 1st/2nd (663) | Other (266) | Positive (675) | Neg. Neutral (330) | Gen Mech Prod (94) | Electric Electron (117) | Aero (56) | Other (122) | Math Physics (92) | Biology Chem (109) | Computers (77) | Bus, Econ, Accts (99) | Social Studies (80) | Humanities Law (136) |
| Finding work intellectually stimulating | 72 | 76 | 63 | 73 | 70 | 72 | 75 | 75 | 66 | 68 | 73 | 77 | 66 | 73 | 79 |
| Working with compatible people | 60 | 62 | 58 | 61 | 57 | 62 | 59 | 61 | 65 | 60 | 63 | 57 | 59 | 53 | 60 |
| Having a friendly management style | 24 | 24 | 26 | 28 | 18 | 36 | 25 | 25 | 30 | 21 | 17 | 29 | 30 | 18 | 15 |
| Feeling that one was doing a good job for society generally | 22 | 19 | 29 | 15 | 37 | 10 | 15 | 11 | 20 | 22 | 28 | 10 | 23 | 40 | 31 |
| Good working conditions | 21 | 19 | 23 | 22 | 18 | 20 | 25 | 29 | 18 | 29 | 19 | 23 | 18 | 18 | 15 |

SOURCE: Opinion Research and Communications, 1985.

**TABLE 6** Quality of Education/Industry Links

|  | Group Trng Associations | Engineering Companies | Large Engrg Companies | Large Companies |
|---|---|---|---|---|
| **Primary Schools** | | | | |
| Excellent | 6 | 5 | 10 | 0 |
| Good | 7 | 13 | 21 | 15 |
| Indifferent | 13 | 14 | 21 | 23 |
| Poor | 13 | 7 | 12 | 12 |
| Non-existent | 54 | 54 | 27 | 38 |
| Not answered | 8 | 7 | 9 | 12 |
| **Secondary Schools** | | | | |
| Excellent | 17 | 19 | 33 | 0 |
| Good | 61 | 44 | 53 | 65 |
| Indifferent | 13 | 13 | 9 | 12 |
| Poor | 3 | 6 | 0 | 4 |
| Non-existent | 3 | 14 | 2 | 12 |
| Not answered | 4 | 3 | 2 | 8 |
| **FE and 6th From Colleges** | | | | |
| Excellent | 17 | 16 | 30 | 4 |
| Good | 44 | 37 | 40 | 46 |
| Indifferent | 17 | 17 | 16 | 19 |
| Poor | 4 | 4 | 1 | 12 |
| Non-existent | 8 | 18 | 7 | 12 |
| Not answered | 10 | 8 | 6 | 8 |
| **Uni and Poly Departments** | | | | |
| Excellent | 1 | 14 | 23 | 15 |
| Good | 26 | 35 | 56 | 77 |
| Indifferent | 15 | 16 | 15 | 42 |
| Poor | 17 | 9 | 0 | 0 |
| Non-existent | 31 | 22 | 5 | 0 |
| Not answered | 10 | 3 | 1 | 4 |
| **Uni and Poly Careers Service** | | | | |
| Excellent | 0 | 4 | 11 | 19 |
| Good | 14 | 19 | 49 | 58 |
| Indifferent | 11 | 19 | 23 | 15 |
| Poor | 17 | 9 | 5 | 4 |
| Non-existent | 46 | 41 | 7 | 0 |
| Not answered | 13 | 9 | 4 | 4 |
| **LA Careers Service** | | | | |
| Excellent | 39 | 7 | 15 | 0 |
| Good | 50 | 29 | 44 | 38 |
| Indifferent | 8 | 22 | 21 | 31 |
| Poor | 1 | 13 | 15 | 4 |
| Non-existent | 0 | 22 | 4 | 19 |
| Not answered | 1 | 7 | 1 | 8 |
| N= | 72 | 209 | 81 | 26 |

SOURCE: Connor, 1992.

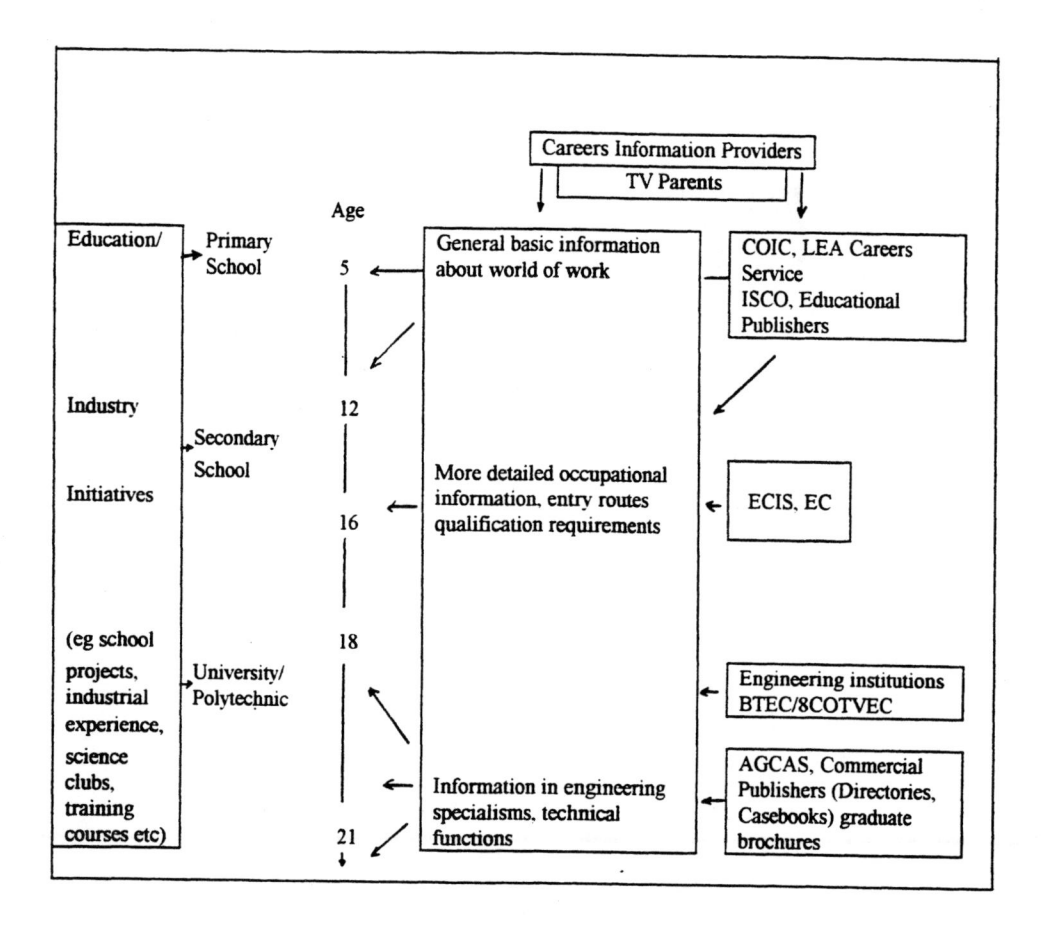

FIGURE 16 Overview of how occupational information on engineering is provided in careers, education, and guidance.

SOURCE: Connor, 1992.

an individual selecting a career in a related area.

As we noted earlier in this paper, individual freedom of choice is a paramount concern in a free market economy. Success is therefore variable. It is also highly dependent upon wider socioeconomic factors. At times of recession, individuals are only too keen to take up whatever employment is available; thus, temporal long run comparisons of the effectiveness of policy are difficult. Similarly, S&T innovatory diffusion is itself a subject of evolution and change. New areas emerge—the biological sciences, information technology—and are presently preeminent, and this can in many subtle ways influence interest and the propensity for uptake of a particular career. Much of industry is no longer the old steam industry of the past; it is more dynamic, cleaner, more high-tech. This again can, *if transmitted to potential applicants*, significantly influence interest and involvement.

Similarly, the perception of academe as an ideal location for a scientific research career has in recent years become somewhat compromised. This is due to a range of resource allocation reasons, which can have a carry-over effect into the expectancy and career search pattern of individuals. Across the EC there is at present only a preliminary stage of cross-national exchange of employment opportunity. Much may change in the post-1992 years and needs to be carefully considered and monitored. (Harmonization of qualifications, standards of employment, and much else will then influence the subject matter we are discussing here, not of least importance will be linguistic ability.)

The policies and mechanisms indicated in Table 7 are not exhaustive, merely illustrative. Some are of a general nature, others are specific to particular cohort

**TABLE 7** Examples of Policies and Mechanisms for Improving the Propensity to Take Up an S&T Career

| Level or Area of Application | Direct or Indirect Effect (D, I) | Level or Area of Application | Direct or Indirect Effect (D, I) |
|---|---|---|---|
| **I  Early School Experience** | | **V  Infrastructural Level** | |
| Improved instruction technique | I | Increased resource to Academe | I |
| Resource support | I | Increased research facilities | I |
| More challenging and involving syllabus | I | Distance Learning programs | I (D) |
| Emphasis upon creativity in Science | I | Involvement in Science Parks | |
| Social societal and environmental awareness | I | Industrial-Academe Campus Linkage | |
| Overcoming too early specialization approach | I (D) | Opening Windows for Engineering | |
| Non-formal institute inputs (e.g., OSC) | I | Neighborhood Engineers schemes | |
| | | Young Engineer for Britain | |
| **II  College Level** | | Local Career Exhibitions, National Exhibitions | |
| Sandwich course schemes | D | SCIP | |
| Industrial Placement Program | D | Insight into Industry | |
| Wider interdisciplinary syllabus | I | | |
| Academic-industrial linkage programs | D | **VI  Demographic or Sociocultural Level** | |
| Industrial sponsorship (research) programs | D | Awareness and understanding | |
| Industrial scholarship program | D | of Science programs | I |
| Generally improved financing | I | Programs specifically geared to | |
| Preferential grants to Science students | D | increasing women's participation (WISE) | D |
| Preferential financial input to | | WISE [Women into Science and | |
| faculty at Institute level | D | Engineering (1984)] | I,D |
| (Unit of Resource) | | Engineering award scheme | |
| HITECC (1988) | | for women (1980-82) | |
| IT Initiative (1983) | | Inright | |
| Engineering Technology Program (ETP) (1986) | | GATE (1984) | |
| Manufacturing Systems | | TESS (1985) | |
| Engineering (MSE) Initiative (1988) | | | |
| | | **VII  International Level** | |
| **III  Postgraduate Level** | | International exchange program | |
| Academic-industrial liaison schemes | D | Comett | I,D |
| Teaching Company Scheme | D | Erasmus | I,D |
| Industrial Sponsorship | D | Petra | |
| Society relevant research projects | I, D | Force | |
| SISCON program | I, D | Eurotecnet | |
| Master of Engineering courses (MEng) | D | (Lingus) | |
| Jupiter program (1989) | I, D | Tempus | |
| Postgraduate Engineering | | (Youth for Europe) | |
| Summer Schools (SERC) | D | Specific Academe Exchange programs; | |
| | | intercollated experience, postgraduate courses; | |
| **IV  Continuation Level** | | Research Council and other Foundation funding | |
| Transfer and reorientation program | D | programs | |
| Integrated Graduate Development program | D | UNESCO Fellowships | |
| Flexible reentry program | D | | |
| Modular retraining program | D | | |

groups. We have attempted to group or cluster the various mechanisms, policies, or concerns in accordance with the decision-schemata referred to earlier. It is important to recognize that this is this writer's classification. There is no national or pan-European body that has carefully thought out a master blueprint. In that sense there may be an element of artificiality. Similarly, the table does *not* imply any degree of individual weighing of the various sub-components, or their relative importance, to the overall scheme of things. It is important to recognize this absence of priorities, for without a blueprint much policy is piecemeal. Policy may thus serve a particular direction but relative importance and resource weighing (and hence adequacy of that weighing) is often not forthcoming. This is an important consideration for further research in this area. For example, who can say, at present, whether resource allocation should be primarily directed at the early stages of the educational process, or at later stages? How can we measure policy efficiency and effectiveness? Many measures come to mind (numbers staying with science, evaluatory scores, attitude measures), but at present they are *not* being explored in a coherent manner. There is much room for a systems-analysis type research procedure here.

Thus, before examining Table 7 in any detail, it is worth making the following overarching point that should influence future analysis. Let us consider the following question: Where is it most beneficial, cost-effective, to apply policies that aim to increase an individuals' propensity to take up an S&T career? More than that, in macro terms, what is the aggregate gain (viz. when we summate across the number of individuals involved)? Have we any idea, any way, of measuring the total societal return? More subtly, are there surrogate influences on which we should act or apply policy levers? For example, if science is better taught, more fully comprehended, does this increase the individual's propensity to stay with the subject to the point of full-time career involvement? Or—much more difficult to analyze—if syllabuses can be evolved that provide a flexible generic base to an individual's understanding and range of applicability of that knowledge, does this then overcome the problem of rigid stratification into particular occupational areas? This is an important question since skill substitution effects can offset more restrictive limited manpower provision. Such questions have been barely explored.

Sooner or later societies will have to examine such questions in fine detail. On the one hand, the high cost of education has been recognized—in most nations it is the higher proportion of state expenditure. On the other hand, the socioeconomic opportunity costs of an insufficiently well-educated, non-motivated, or insufficiently scientifically and technologically involved populace is a major limit, perhaps *the* limiting factor on socioeconomic progress. However, because of the high input cost at the aggregate level, we may be forced to examine policy mechanisms in terms of a wider societal value-added approach. (See Figure 17.)

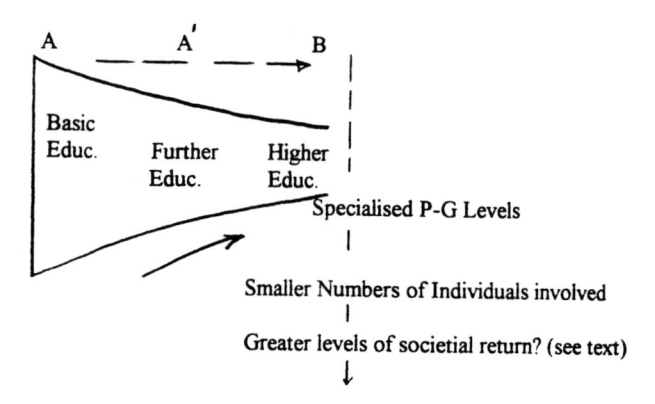

**FIGURE 17** Selective filtering.

At level A, basic education (which involves the mass of society), we could seek, by various policies and research inputs, to improve instruction and involvement. The cost would be high, but it would touch hundreds of millions of individuals. If this induced, say, 10-20 percent more individuals to *want* to seek higher education, to follow through into S&T careers, the social return is equally enormous.

Alternatively (or in addition), we might wish to concentrate effort at level A' or B. Smaller numbers of individuals are involved, but because of the degree of training (sunk costs), the value added of not losing them from S&T careers at this stage justifies particular policy input. We will not explore all the possibilities here, or provide a detailed socioeconomic return calculus, other than to make the point that the shopping list portfolio of policies inherent in Table 7, which reflects in many ways eclectic and incoherent state policy, might benefit from a more holistic socioeconomic analysis. To date this has been insufficiently addressed. *It is a major research requirement.*

Having made this general analytic point, if we examine Table 7 we can observe that policies are being applied at most levels of the educative process and at many industrial-educational, demographic-educational, and societal-educational interface points. Inter alia, they may have considerable direct, or indirect, effects upon both the propensity for greater numbers of individuals to follow an S&T career and for them to be more productive having done so. It would be overclaiming, however, that all of the programs or policies listed in Table 7 are *specifically* aimed at increasing the propensity to choose an S&T career. While some are (e.g., those geared to increasing the participation of women in science, engineering, etc., or industrial sponsorship programs aimed at engineering undergraduates), many of the other policy inputs listed in Table 7 may well have just as great an effect in terms of influencing an individual's aspirations to undertake an S&T career.

A few examples follow: a syllabus design that is less theoretical in emphasis, more explorative, can do much to motivate and involve, and improved teaching techniques can get individuals across the decision threshold (or certification-grade requirement) to participate in higher education (on a science course). Once over *that* hurdle, the chances are greatly increased that an S&T career may be followed. Or, in another attitudinal area, close industrial-academic liaison that demonstrates the challenges to be found in industry (similarly with increased attention to socio-environmental issues, say) can lead to the *desire* to be involved. Thus, in many ways, the cognitive and affective dimensions intertwine.

Two guiding principles might be considered for the formulation of policy geared to increase the likelihood of later propensity to take up an S&T career. They are as follows: (1) the greater the number of individuals who progress with a scientific and technological education (across the various decision thresholds—see Figures 1 and 7) then, in statistical terms, the greater the chances of later career involvement; and (2) the opportunity cost to society of individuals who have progressed right across the various educational levels and do not *then* take up an S&T career is considerable. Thus, in many ways, these two somewhat opposing principles have to be influential in shaping policy deliberations. The extent to which this is, at present, the case is debatable.

Also, we must consider the external world. Career uptake, as we have seen, depends also upon perceived image of industry, of academe's (teaching and research) condition, and of *relative* social, economic, and psychological reward. To a large degree this external world is not subject to change via centralized policy. Industry has to sell itself, persuade, cajole. Every day, through marketing and the competitive stimulus, it does this with respect to its products: the skill is to do this with regard to its own organizational and cultural condition.

Finally, it should be noted that this paper has not considered in any detail the demand side of S&T careers (or, perhaps of even more importance, the *changing* pattern of demand: new skill requirements, new combinations of skills). Undoubtedly, demand has an effect upon choice of career, as do *new* opportunities, new challenges. The *interfacing* in policy and analysis terms of supply and demand in S&T occupation no doubt merits a full, and later, analysis. Necessarily, any such analysis (and subsequent research programs) should spill over into the influence of changing demand, changing S&T structure, into educational policy *at all* levels of the education process.

We have also noted that there are many difficulties of coupling, of sufficiency of knowledge between student, graduate, or academe on the one hand, and the real world of scientist, industry, or research institution on the other. The opportunity to mix, meet, intermingle prior to *various stages* of the decisionmaking process regarding career choice is obviously of much potential and actual importance. At graduate (and international) levels, such EC programs as Erasmus or Comett are of much importance and relevance.

The principal aims of Erasmus are "to increase the number of students spending a period of study in another member state, to foster cooperation among universities in all member states, and to increase the mobility of teaching staff and thus improve the quality of education and training provided." With respect to Comett, "the objective is to improve technological training, especially in the advanced technologies, the development of highly qualified human resources and hence the competitiveness of European industry."

The program comprises three elements: (1) cooperation between industry and universities, (2) better training in new technologies, and (3) cross-border cooperation. More specifically, with regard to Comett, there are five main strands of activity:

*Strand A*—development of University Enterprise Training Partnerships

*Strand Ba*—student placements in enterprises

*Strand Bb*—fellowships

*Strand C*—development and testing of joint university-enterprise projects in continuing education

*Strand D*—multilateral initiatives for the development of multi-media training schemes.

Both directly and indirectly, such programs facilitate knowledge, awareness, and challenge in the S&T area. They can, therefore, greatly upgrade the propensity to take up an S&T career. The Comett program has been evaluated by Whiston and Senker (1989) in a large study commissioned by the EC (jointly undertaken by SPRU, University of Sussex, and Coopers and Lybrand). The program was seen to be working well. Many projects were identified that would not have been undertaken without some EC sponsorship. The quantitative effect in terms of later career uptake (or attend career aspirations) was neither sought in that study, nor known. It is not difficult to argue, however, that such important, large-scale international initiatives play an important stimulatory role in encouraging later careers in S&T.

Undoubtedly, the opportunity to communicate and work with peers and to experience new S&T challenges in new settings provide such stimulation and encouragement. Both Erasmus and Comett are important facilitators and bridging mechanisms in that respect.

In EC terms, in addition to these two programs, we would also note the value of the following:

- *Petra*, which seeks to ensure vocational training;

- *FORCE*, which addresses training schemes, vocational training;

- *Eurotecnet*, which considers both "basic and continuing vocational training, with the aim of taking account of current and future technological change and its impact on jobs and work and the qualifications and skills that are needed";

- *Lingua*, which promotes language improvement;

- *Tempus*, which provides EC assistance in restructuring HE in the countries of Central and Eastern Europe in order to facilitate their speedy adaptation to the requirements of the market economy; and

- *Youth for Europe*, which promotes the exchange of young people (outside school) and encourages awareness of Europe.

The extent to which specific S&T components permeate through such programs varies considerably. However, the coupling of academe and the wider society is a common component. In some programs (viz. Erasmus and Comett) specific exchange, opportunity of involvement with other scientists, is a large part of the program.

Individual countries have their own initiatives (i.e., the United Kingdom: Teaching Company Scheme, CRAC, IGD, industrial scholarship programs; the United States: "Learning from Scientists at Work," etc.). Evaluation and analysis of such programs from the perspective of influence on career choice would be invaluable as a further guide to policy formulation.

This takes us to our last comment. It is a tentative one, but worthy of careful consideration. If one scans the more contemporary literature regarding S&T career choice, it is immediately obvious that much research focuses upon the male-female gender participation issues (and also upon minority groups). In one sense what we may be seeing here is an attempt to expand the demographic catchment area beyond its historical pattern. That in itself is worthy and very necessary. At the same time, much literature testifies to the large proportion of individuals who, having received much expensive S&T training, decline to take up a career (hence, the pressure to expand the social-catchment area). *Much* further policy research is required on these latter topics to ensure the most effective and socioeconomically beneficial returns for both the individual and for society-at-large. Broad-brush policies, initiatives based on faith, may well have to be more finely attenuated as the scale and form of need becomes more critical to the development of society. In one sense we are talking of areas that account for *billions* of dollars of expenditure, of high wastage-rates dependent upon one's perspective. We are also seeing critical national and world sociotechnical needs partially unfulfilled due to human resource bottlenecks, which demands an even wider policy analysis or research agenda than has been signaled in this brief paper.

# FUTURE RESEARCH TASKS

What now needs to be done, to be taken further? We would suggest the following research tasks, which if *acted upon* in more detail and with greater precision, may help to improve the uptake into S&T career paths:

1. Provide a fully developed *systems model or assessment* of the *whole* educative process in order to guide the relative weighing, importance, and role of each of the present separate policies that obtain at each phase of the educational process. (This is an economic, strategic, evaluatory, and sociotechnical task.)

2. Identify more clearly the *critical individual steps* that influence the life cycle decision pathway and evolution of an individual's experience *prior to* choosing a career, and then assess policy needs.

3. Evaluate the effectiveness of all policies more fully through wide and detailed comparative survey.

4. Examine the major *perceptual* barriers and ignorance that characterize students' knowledge of S&T careers outside academe. Translate this into much greater synergies between academe and commerce, and academe and society at *all* levels of the educative process.

5. The importance of *flexibility* at every facet of education, syllabus, re-entry possibilities, organizational structures, and combination of courses cannot be overemphasized.

Flexibility, however, is a much misused term. It is often a euphemism for uncertainty and ignorance. Therefore, care must be taken in understanding in human resource terms exactly what one is trying to address and achieve. To formulate policy based upon past experience of industry, commerce, and academic structure is dangerous. The need is for detailed analytic study of likely future S&T trajectories, future skill needs, future research priorities. To involve government, industry, commerce, and academe in that analysis and later transmission to the whole student body by the widest means possible. Organizational synergy comes with informational and perceptual transparency. Fluidity of movement improves with individuals who are well informed and confident of their knowledge. We see, there, a massive information distribution task.

We have seen the difficulties of manpower forecasting; nevertheless, the social costs of error are enormous. Equally, the cost of training numerous individuals only to see them not take up S&T careers presents problems (though we must recognize that they contribute to the general economy in other areas).

The general concerns signaled in this paper can be translated into the need for more data and surveys in the areas listed below in which more detailed national and *international* comparative data are required—especially at a time of increasing migratory and exchange programs and new national linkages at the global level. This requires improved knowledge regarding the following questions:

- What *major factors* influence career choice in S&T and how might these be changing under new world conditions?

- Can we *rank* these influences into differing orders of magnitude?

- What are the major *career decision steps* and how might they be influenced by policy enaction? What proportion of individuals are lost at each stage?

- Are there different patterns relating to *subject* area (e.g., physical sciences, engineering, biological sciences, computing) and also career choice?

- How is all of this influenced by an individual's educational/training background?

- What are the major *barriers* and obstacles to the likelihood of taking up a certain career pathway? How does this vary from subject area, discipline, and occupational area?

- How is career choice influenced by the changing S&T scene, changing world conditions?

- What *policies* are most effective in encouraging *placement* in (or *return to*) an S&T career?

- What are the effects of oversupply (nationally and internationally) in discipline areas?

- What differences are there relating to career choice regionally, nationally, and culturally? Are there significant cultural differences?

- Are there different patterns and mechanisms relating to career uptake at different *levels* (e.g., technician, undergraduate, and postgraduate)?

- In what ways is the process influenced by the *eliteness* of the educational establishment or university or college?

- *Which* S&T career paths are the most (or least) *sought after* and why? How does this vary from nation to nation?

- What different patterns and mechanisms of choice are observed between *gender* or *ethnic* grouping?

- Are there significant national or international *poaching* problems across careers or nations? Which *non*-S&T areas lay the biggest claim to S&T graduates?

- What *international/governmental developments* are playing a significant role in our understanding of the above (e.g., international migration ease, standard qualifications and training, multinational linkage, etc.)?

- *Which* S&T *subjects* or disciplines lead to the highest uptake in S&T careers? How does this vary internationally?

- Does the degree of *specialization* (or multidisciplinary training) play a significant role in subsequent career uptake?

At present we lack sufficient data or knowledge in all the above areas. This compromises our ability to formulate the most effective policy agenda with respect to influencing career choice in S&T areas. The subsequent loss to society at large, to scientific endeavor, and to technological and industrial development is no doubt considerable.

## NOTES

1. In this sense, much of the research and extensive writings of my old mentor, Professor John Cohen (Manchester University, UK), in contradistinction to such researchers as Eysenck or Cattell, might be usefully considered.

2. See Whiston, T.G. (1969), Future Trends in Science Education, *Education in Chemistry* 6(4), pp. 133-136.

3. In using the term "career," we are not distinguishing the type of occupation, which may be in the public or private sector, industry or academe, productive, research, teaching, or whatever. Obviously, perceived opportunity, market conditions, individual interest, and capability all play a part.

4. The recent removal of the HE "binary line"—separating polytechnics and universities—is also of much importance to our analysis here. The polytechnics (now "universities") have a tradition of applied work, of graduates seeking S&T careers. In future years, UK statistics may therefore change significantly.

# REFERENCES

Astin, A. 1991. The Changing American College Student: Implications for Educational Policy and Practice. Higher Education. 22:129-143.

Atkinson, J. 1989. Corporate Employment Policies for the Single European Market, IMS Report No. 179, IMS University of Sussex.

Bee, M. and P.G. Dolton. 1989. What Do Graduates Earn? The Starting Salaries and Earning Prospects for University Graduates 1960-86. University of Bristol, Department of Economic Working Paper 89/246.

Bee, M and P.J. Dolton. 1990. Where Do Graduates Go? The First Destinations of University Graduates, 1961/62-1986/87. Studies in Higher Education. 15(3)313-329.

Benditt, J. 1992. Women in Science: First Annual Survey. Science. 255(5050):1363-88.

Birch, W. 1988. The Challenge of Higher Education: Reconciling Responsibilities to Scholarship and to Society. SHRE/Open U.P.

Bon, C.A. et al. 1991. Effects of a Pharmacy Student Recruitment Programme. American Journal of Pharmaceutical Education. 55(3):258-63.

Boys, C.J. and Kirkland, J. 1988. Degrees of Success: Career Aspirations and Destinations of College, University and Polytechnic Graduates. London: Jessica Kingsley Publishers.

Brennan, J. 1992. European Higher Education Systems: Germany, the Netherlands, the UK. CNAA.

Clark, B. (ed). 1987. The Academic Profession: National, Disciplinary and Institutional Settings. Berkeley: University of California Press.

Confederation of British Industry. 1989. Towards a Skills Revolution. Report of the vocational education and training task force. London CBI.

Connor, H., C. Jackson, G. Pike, and B. Ball. 1992. A Ccareers Service for Engineering, IMS Report No. 219 IMS University of Sussex.

Corporation for Public Broadcasting, Washington, D.C. [Sponsoring Agency] Science Lives: Women and Minorities in the Sciences. Minnesota University, Minneapolis, KUOM Radio (13 half hour programmes).

Council for Industry and Higher Education. 1987. Towards a Partnership: Higher Education, Government, Industry.

Daune-Richard, A.M. 1992. Women in "Male" Careers: The Case of Higher Technicians in France. Centre d'Etudes et de Recherche sur les Qualifications, Paris (France). Training and Employment: French Dimensions Newsletter 6:1-4.

Davis, J.A. 1965. Undergraduate Career Decisions, Aldine Pub. Co. Chicago DES (1975), Curriculum Differences for Boys and Girls, DES Education Survey 21, London HMSO.

DES. 1991. Education and Training for the 21st Century, HMSO.

Dick, T.P. and S.F. Rallis. 1991. Factors and Influences on High School Student's Career Choices. Journal for Research in Mathematics Education, 22(4):281-92.

Dolton, P.J. 1989. The Early Careers of 1980 Graduates: Work Histories, Job Tenure, Career Mobility and Occupational Choice. Department of Employment Research Paper.

Durndell, A. 1990. Why do Female Students Tend to Avoid Computer Studies? Research in Science and Technological Education, 8(2):163-70.

Eden, D. 1992. Female Engineers: Their Career Socialization in a Male-Dominated Occupation. Urban Education, 27(2):174-195.

Engineering Professors Conference Research Unit. 1986. Careers of Scientists and Engineers. University of Liverpool.

Eurostat. 1992. Europe in Figures, EEC (3rd Ed.) Brussels.

Finniston, M. 1980. The Finniston Report. Engineering Our Future: Report of the Committee of Inquiry into the Engineering Profession. London HMSO Cmnd 7794.

First Destination Statistics, USR/CDO, Annual Report.

Freeman, R.B. 1971. The Market for College-Training Manpower: A Sstudy in the Economics of Career Choice. Cambridge: Harvard University Press.

Frieden, C. and B. Fox. 1991. Career Choices of Graduates from Washington University's Medical Training Program. Academic Medicine, 66(3):162-64.

Fuller, A. 1991. There's more to science and skills shortages than demography and economics: attitudes to science and technology degrees and careers. Studies in Higher Education, 16(3):33-341.

Fulton, O., A. Gordan, and G. Williams. 1982. Higher education and manpower planning: A comparative study of planned and market economies. Geneva: International Labour Office.

Galbraith, M. 1992. Understanding career choices in men in elementary education. Journal of Educational Research, 85(4):246-253.

Garrett, L. 1986. Gender differences in relation to science choice at "A" level. Education Review, 38(1).

Gellert, G. (ed). 1992. Higher education in Europe, J. Kingsley.

Gordan, A. and Pearson, R. 1984. Manpower for IT. IMS Report No. 83. A report to SERC, IMS University of Sussex.

Gordan, A., R. Hutt, and R. Pearson. 1985. Employer sponsorship of undergraduate engineers, IMS April 1985.

Gothard, W.P. 1982. Brightest and the best: A study of graduates and their occupational choices in the 1970s. Chester: Bemrose Press.

Hackett, E.J. 1990. Science as a vocation in the 1990s: The changing organizational culture of academic science. Journal of Higher Education, 61(3):241-279.

Higher Education Policy (Special Theme). 1992. Research and Training: Towards innovative strategies of financing. The Quarterly Journal of the International Association of Universities. 5(2).

Hilska, U. 1990. Recruitment of students to Universities of Technology in Finland. European Journal of Engineering Education, 15(4):361-368.

HMSO. 1987. Higher Education: Meeting the Challenge, Cmnd 114 1987. House of Lords Select Committee on Science and Technology (1991) Innovation in Manufacturing Industry, Vol. 1 - Report HMSO London.

Imperial Ventures Ltd. 1989. Survey into the attitudes and intentions of graduating engineers.

Kelly, A. (ed). 1981. The missing half - girls and science education. Manchester: Manchester University Press.

Kelly, A. 1988. Ethnic differences in science choice, attitudes and achievement in Britain. British Educational Research Journal, 14(2).

Kelsall, R.K. 1957. Report on an inquiry into applications for admission to universities. Committee of Vice-Chancellors and Principals of the Universities of the U.K. Published by the Assoc. of Universities of the British Commonwealth. London.

Kelsall, R.K., A. Poole, and A. Kuhn. 1972. Graduates: The sociology of an elite. Methven, London.

Kennan and Newton. 1979. A Longitudinal study of engineers. University of Edinburgh. Kennedy, A., Rotherham, L. and Whiston, T. G. (1983) A review of the SERC Integrated Graduate Development Programme. 2 vols. 1983 [published by Technical Change Centre London TCCR-83-005 Report to SERC]

Lawton, D. 1973. Social change, educational theory and curriculum planning. London: University of London Press.

Lent, R.W. 1991. Mathematics self-efficacy: Sources and relation to Science-Based careers choice. Journal of Counselling Psychology, 38(4):424-430.

Lips, H.M. 1992. Gender and science-related attitudes as predictors of College student's Academic Choices. Journal of Vocational Behaviour, 40(1):62-81.

Luttikholt, H.W. 1989. EED Sources of support for higher education and research in 1990: A policy paper. Liaison Committee of Rectors Conferences of Member States of the EC, Brussels 1989.

Mabey, C. 1986. Graduates into industry - a survey of changing graduate attitudes. Henley Management College.

McCormick, K. 1988. Engineering education in Britain and Japan: Some reflections in the use of 'Best Practice models' in International Comparison. Sociology, 22(4):583-605.

Mervis, J. 1991. Educators rally to salvage science dropouts. Scientist, 5(4) p 1,4,10. Morgans, C.J. (1991) Engineering the way to becoming a Federal Engineer. Journal of Career Planning and Employment, 52(1):38-42.

National Science Foundation. 1990. NSF's Research experiences for Undergraduates (REU) Program: An assessment of the First three years NSF Report 90-58 29 pp.

National Academy of Sciences, Office of Scientific and Engineering Personnel. 1991. Women in Science and Engineering: Increasing their numbers in the 1990s - A statement on policy and strategy. Washington, D.C: National Academy Press.

NFER (for DES) Noncompletion of engineering courses (ongoing).

NFER (for DES) Images of Engineering (ongoing, 1990).

Opinion Research and Communication. 1985. Attracting the brightest students into industry. Vol. 2. Career attitudes and aspirations of the undergraduate population. London, August 1985.

OTA. 1991. Federally funded research: Decisions for a decade [Chapter 7: Human sources for the research work force]. Congress of the United States. Washington DC OTA-SET-490.

Otto, P.B. 1991. One science, one sex? School science and mathematics, 91(8):367-372.

Parsons, D. 1986. Employment Brief No. 8 - School Leaver Supply Trends for the 1980s, IMS, University of Sussex.

Pearson, R. 1989. Postgraduate training and careers in the United States—Trends and Issues. Report to SERC, IMS University of Sussex, IMS Report No. 165.

Pearson, R. et al. 1990. The recruitment and retention of University academic and academic related staff. Report to the Assoc. of University teachers and CVCP. Sept. 1990, IMS University of Sussex.

Pearson, R., G. Pike, and S. Holly. 1991. The IMS Graduate Review 1991, IMS Report No. 206, IMS University of Sussex.

Pearson, R., F. Andrentti, and S. Holly. 1990. *The European Labour Market Review: The key indicators.* IMS Report No. 193, IMS University of Sussex.

Pearson, R. and G. Pike. 1989. The graduate labor market in the 1990s. IMS Report No. 167, IMS University of Sussex.

Pearson, R. 1988. Scientific Research Manpower: A review of supply and demand trends. IMS.

Pearson, R., G. Pike, A. Gordan, and C. Weyman. 1989. How many graduates in the 21st Century?—The choice is yours. IMS Report No. 177, IMS, University of Sussex.

Pike, G. and H. Connor. 1990 Evaluation of the Teaching Company Scheme—The Associates. IMS Paper No. 159 IMS University of Sussex.

Post, P. 1991. Self-efficacy, interest and consideration of math/science and non math/science occupations among Black Freshmen. Journal of Vocational Behaviour, 38(2):179-186.

Prais, S. J. 1989. Qualified manpower on engineering: Britain and other industrially advanced countries. National Institute Economic Review, 127:76-83.

Raisman, J. 1991. Industry and Education: Strengthening the partnership. Paper to the AGM of IMS. IMS Paper No. 168, December 91, IMS University of Sussex.

Redpath, B. and B. Harvey. 1987. Young people's intentions to enter Higher Education, CS/HMSO London 1987.

Reid, M. I., B.R. Barnett, and H.A. Rosenburg. 1974. *A matter of choice.* A study of guidance and subject options, Slough, NFER.

Research Surveys of Great Britain Ltd. 1990. European students attitudes to business, Report No. NJ 5896.

Ryrie, A.C., A. Furst, and M. Lauder. 1979. Choices and chances: A study of pupils' subject choices and future career intentions. Scottish Council for Research in Education. Hodder and Stoughton, London.

Sanyal, B.C. 1987. Higher education and employment: an international comparative analysis. Lewes: Falmer Press.

Sax, L. 1992. Predicting persistence of science career aspirations: A comparative study of male and female college students. [Paper presented at the America Educational Research Association Conference, San Francisco, CA; April 24, 1991.]

Senker, P. 1992. Industrial training in a cold climate: Assessment of Britain's training policies, Aldershot, Avebury.

Smith, J. 1987. Graduate recruitment to Engineering. EITB 1987.

Stock, J., R. Pearson, and H. Connor. 1990. The careers and training of engineers—some research issues. IMS, University of Sussex.

Stock, J., H. Connor, and R. Pearson. 1990. The careers and training of engineers. SRC/IMS, IMS University of Sussex.

Taber, K. S. 1992. Science-relatedness and gender-appropriateness of careers: Some pupil perceptions. Research in Science and Technological Education, 10(1):105-15.

Taylor, J. 1985. Employability of graduates: Differences between universities. University of Lancaster.

THES. 1992. V-cs fear 'extreme' cuts. The Times Higher Educ. December 4, 1992 p 3. (S. Griffiths and C. Sanders).

Vaughan, D. K. 1990. The image of the Engineer in the Popular Imagination, 1880-1980 Bulletin of Science, Technology and Society, 10(5-6):301-304.

Waite, R. and G. Pike. 1989. School leaver decline and effective local solutions. IMS Report No. 178. IMS University of Sussex.

Walker, B. 1991. Environmental health and African Americans. American Journal of Public Health, 81(11): 1395-1398.

Waltner, J. C. 1992. Learning from scientists at work. Educational Leadership, 49(6):48-52.

Wankowski, J. 1968. Random sample analysis: Motives and goals in university studies. Educational Survey. University of Birmingham.

Warren-Piper, D. and S. Acker (eds). 1984. Is Higher Education fair to women? SRHE 1984.

Weir, A. D. and F. Nolan. 1977. Glad to be out? Edinburgh, Scottish Council for Research in Education.

Whiston, T. G. 1988. The coordination of education policies and plans with those for Science and Technology: Western Europe and Developing Countries [plus an annotated bibliography of 400 references]. Complete Issue of The Bulletin of the International Bureau of Education. April-June 1988, 247:1-144. Geneva. (In English and French).

Whiston, T. G. 1987. Restructuring and selectivity in academic science. Science Policy Support Group (SPSG) London. Concept Paper No. 5.

Whiston, T. G. (ed). 1979. The uses and abuses of forecasting. London: Macmillan.

Whiston, T. G. 1992. Managerial and organizational integration. London and Berlin: Springer-Verlag.

Whiston, T. G. and J. Senker. 1989. Evaluation of the COMETT Programme, Brussels: Commission of the European Community April 1989. (Also available in German and French).

Whiston, T. G. 1987. The training and circumstances of the Engineer in the Twenty-First Century. A study undertaken for the Fellowship of Engineering, March 1987, SPRU University of Sussex, 234 pp.

Whiston, T. G. 1984. Written evidence on New Technologies and UK Educational Policy Response. House of Lords Select Committee on Science and Technology: Education and Training for new technologies. HMSO 3:472-478, London.

Whiston, T. G. 1969. Future trends in science education. In Education in Chemistry (Journal of the Royal Institute of Chemistry), 1969 6(4) 133-136.

Whiston, T. G. and R. Geiger (eds). 1992. Research and Higher Education: The UK and USA Scene. Open University Press Milton Keynes.

Whiston, T. G. 1986. Management and assessment of interdisciplinary training and research, Paris: ICSU and UNESCO Press.

Whiston, T. G. 1982. Interdisciplinary Training: Problems and perspective—a study of the ESRC-SERC Joint Research Committee Doctoral Programme (3 vols.). Report to ESRC-SERC. University of Sussex 1982 (Also published in condensed form by SERC as Interdisciplinary Research: Selection, Supervision and Training - a summary of the Whiston Report. SERC Monograph April 1983, Swindon.

Whiston, T. G., P. Senker, and P. MacDonald. 1980 An annotated bibliography on the relationship between technological change and educational development. Paris: UNESCO Press.

White, R. M. 1991. Engineer your child's way to the top, PTA Today, 17(1):18-19.

Women in Science Newsletter. 1992. [Issued by SPSG, London] updating details of WITEC (Women in Technology in the EC) and WISE (Women into Science and Engineering launched by the Engineering Council and the Equal Opportunities Commission).

Woods, P. 1976. The myth of subject choice. British Journal of Sociology, 27(2):130-149.

# Factors Behind Choice of Advanced Studies and Careers in Science and Technology: A Synthesis of Research in Science Education

## Torsten Husén

In 1989, as part of the celebration of its 250th anniversary, the Royal Swedish Academy of Sciences held an international symposium on issues in science education. The subtitle of the symposium report was *Science Competence in a Social and Ecological Context* (Husén and Keeves, eds., 1991). Since I was invited to organize the symposium, I shall take some of the issues we dealt with as a point of reference for the present paper. The main concern behind our present conference is the difficulty encountered by many industrialized countries on both sides of the Atlantic in recruiting a sufficient number of young people to careers in science and technology. It is a problem that has to be seen in a wider context than we usually do. It is not just a problem of getting enough young people with good specific abilities and competencies. In our highly complex, technological society we need to recognize the strong motivating effect of the growing concern of the ecological and social effects of applied sciences and how they affect the public image of science and its uses. Thus, science education reaches far beyond the pedagogical problem of what abilities are needed, what specific competencies should be taught to young people, and what standard of knowledge should be achieved.

## BRIEF REVIEW OF RELEVANT FACTORS

What factors have an impact on young people's motivation to embark upon advanced studies in the natural sciences and eventually upon their willingness—or reluctance—to pursue careers in the field? I shall briefly review what many would regard as the main factors.

1. The *home background*, particularly parental education and occupation, is important not only in providing role models but also in shaping attitudes and influencing motivation. Mention should be made of the sociocultural milieu with its gender stereotypes and influence on attitudes.

2. I have already mentioned the *image of science* young people get in today's society via the home, the school, and, not to mention, the media. Increasingly over the last few decades, young people have begun to ask themselves to what extent science and its applications are beneficial or harmful to mankind. Science has become integrated into the economy of industrialized societies in what could be called a techno-system. George Henrik von Wright (1989), in his Jubilee Lecture on *Science, Reason, and Value*, pointed out that "a loss of prestige for science due to the misuse made of it in technology . . . (implies) a weakening of the intellectual curiosity which is the psychological motive for the epistemic orientation of science." This tends to affect young people's attitudes toward science as a field of study and their willingness to embark upon careers in science and technology.

3. The *teaching of science* in school, how well it is taught and how difficult and elitist it is perceived by the students, tends to have a profound influence

on the willingness to pursue advanced studies. Attitudes to science are shaped at a very early stage in school. Not in the least, girls are turned away by the elitist perception that science is something mainly for boys, which is closely linked to teacher competence in this particular subject area. There is no need here to point out the difficulties one has in many countries in recruiting *competent teachers*. Those who study science at the tertiary level are attracted by several other, better remunerated careers than teaching. I was struck by this when I was invited to a hearing by the National Commission on Excellence a decade ago. So many high school science teachers in the United States have not even majored in science.

4. Science is often perceived as a *difficult school subject*, particularly by girls. Its close relationship with mathematics and its own abstract features tend to have an abhorrent effect on students. It deals with strict laws, principles, and rules that cannot be substituted by verbal form.

5. Of special concern is the big *gender gap* in the percentage of young people embarking upon advanced studies and careers in science and technology. Girls represent an enormous, untapped reserve of ability in the field, even with regard to existing sex differences (whatever their cause may be) in science achievement in school. I shall devote a major part of the paper to sex differences in science achievement, abilities related to science, and motivational orientation toward science studies and careers.

6. Finally, *career prospects* in the higher education system, and in business and industry, play an important role in attracting young people with a high level of abilities and creative minds. These prospects, in turn, depend upon the financial auspices and the willingness of the political system to invest in science and technology.

In his analysis of factors influencing participation in science after completion of secondary schooling, Keeves (1992) identified the following:

- science values (as reflected in attitude scales measuring career interest and perceived beneficial aspects of science);

- aspirations (expected postsecondary education and subsequent occupation);

- amount of science studies (courses, class time, and homework); and

- science attitudes (interest in science and perceived ease of learning science).

The data analyzed were collected by the International Association for the Evaluation of Educational Achievement (IEA), which conducted two international surveys in some twenty countries. The multi-variate analysis of IEA data presents no evidence that the sex of students has a significant *direct* effect on future participation in advanced science studies or occupations. However, the striking sex differences in participation are due to *indirect* effects of values, attitudes, aspirations, sex stereotypes, and amount of science studies at the secondary school level. These attitudes and values are shaped by the sociocultural environment.

## ATTEMPTS TO PROVIDE EMPIRICAL EVIDENCE

The factors influencing science achievements and aspirations to pursue careers in science shall be elucidated by drawing upon existing empirical data, most of which comes from the IEA cross-national surveys. I do not pretend to present a complete, comprehensive, state-of-the-art picture of the research on what makes young people opt for or not opt for advanced studies and courses in science and technology. My ambition is limited mainly to cross-national surveys in mathematics and science, which I was instrumental in launching in the early 1960s, and which have since been repeated. Thus, I am referring to the science education surveys conducted under the auspices of the IEA.

First, I will provide a thumbnail sketch of the enterprise. A feasibility study was carried out in 1959-1961 (Foshay, ed., 1962), and the first full-scale survey took place in the mid-1960s and targeted mathematics (Husén, ed., 1967). In 1970, the first science study was conducted in 19 countries on students at 3 grade levels in primary and secondary schools (Combes and Keeves, 1973). A second science survey occurred in 1983-1984 in 23 countries (Rosier and Keeves, 1991;

Postlethwaite and Wiley, 1991; Keeves, ed., 1992), and a third survey is being conducted in some 40 countries; data collection on the two occasions is presently being prepared and is expected to be completed by 1999.

It would take me too long even if I tried to deal briefly with the main complicated technical problems we have to cope with in conducting studies of this magnitude. Instead, let me simply hint at some of the problems and refer to the IEA literature, which by now comprises more than 50 monographs in book form and hundreds of minor publications in scholarly journals. Three major technical problems stand out:

1. Selecting *target populations* and drawing *representative samples* have had to be identified at various levels of the national educational systems.

2. *Test instruments* for assessing student competence in science and attitudes toward science and science careers have had to be devised. This requires, among other things, thorough analyses of science curricula in all participating countries and testing the proposed sample questions in each country. Of the latter, very few questions survived the scrutiny needed to make the achievement and attitude tests cross-nationally valid. In addition to the tests, questionnaires had to be devised and administered to students, teachers, and school administrators.

3. *Data processing techniques* and—above all—*multivariate analytic techniques* have to be developed to cope with the enormous mass of data. The analyses were aimed at accounting for the differences between countries, schools, and students in terms of social background factors, teacher competence, methods of instruction, and school resources.

Since IEA has conducted surveys with certain time intervals, it is now possible to measure *trends* over time in science competence and attitudes toward science. The Third International Mathematics and Science Study will be conducted during the 1990s. It entails a *follow-up* component (i.e., the same students are tested with a few years interval). This allows us to assess the growth of competence in the individual student over time and will give us a better picture of the trends than those of cross-sectional studies.

## ATTITUDES AND VALUES AFFECTING PARTICIPATION, ACHIEVEMENT, AND CHOICE OF CAREER IN SCIENCE (IEA SURVEYS)

In the following I shall deal mainly with IEA attitudinal studies focusing specifically on:

1. interest in and attitudes toward science as a *school subject*, and perceptions of its difficulties;

2. perceptions of the *beneficial* and *harmful* aspects, respectively, of science in society; and

3. *career* interests in science.

We shall deal with secondary school students at the beginning and end of the stage (i.e., with 14-year-olds and 18-20-year-olds, respectively).

In the First International Science Survey (FISS) (Combes and Keeves, 1973), students at the 10-, 14-, and 18-20-year-old level were given attitude and interest inventories where they expressed, among other things: (1) interest in science, (2) attitudes toward school science, and (3) attitudes toward the role of science in the world today.

Those in FISS who explicitly disliked science among those *taking* it in the final grade of secondary school were, as one would expect, rather few in comparison with those who did not take it. Those expressing dislike ranged from 22 percent in Germany to 6 percent in Sweden. The percentage among those *not* taking science ranged from 70 percent in England to 14 percent in Hungary.

The responses to the attitude items were combined into a standardized *scale*, Belief in Science, with an international mean of 0. (See Figure 1.) On the positive side, we consistently find Hungary (far above the other countries), and, on the negative side, the Netherlands, Sweden, England, and India, strangely enough, among those also taking science in the last grade of secondary schooling. This was at a time when young people began to realize the ecological effects of applied science.

In the Second International Science Survey, 1983-1984, attitudes toward science and its effects were studied more in depth by means of five scales

**Boys, 14- and 18-year-olds, in Various Study Programs**

| Standardized Mean Score | 14-year-old Students | 18-year-old Non-Science Students | 18-year-old Science Students | Cumulative Distribution For All (%) |
|---|---|---|---|---|
| | | | .91 Hungary | |
| | | .88 Hungary | | |
| .86 | | | | 80% |
| | .67 Hungary | | | 70% |
| .55 | | | | |
| | | | .47 Finland | |
| | .35 Thailand | | | |
| | | | .34 England | |
| | | | .33 Thailand/USA | |
| .27 | | .28 Chile | | 60% |
| | | .20 Finland | | |
| | | .19 Thailand | | |
| | | | .17 Chile | |
| | | | .13 W. Germany | |
| | .12 USA | | | |
| | | | .06 Australia | |
| | | .01 USA | | |
| 0 | | INTERNATIONAL MEAN SCORE | | 50% |
| | -.16 Chile | -.16 W. Germany | | |
| | | | -.17 India | |
| | -.18 Finland | | | |
| | | | -.21 Sweden | |
| | | -.25 Netherlands | | |
| -.27 | -.27 Australia | | | 40% |
| | -.29 Japan | | -.29 Netherlands | |
| | -.35 W. Germany | | | |
| | | -.38 Australia | | |
| | -.40 Sweden | -.40 England | | |
| | -.44 England | | | |
| | | -.47 India | | |
| | | -.48 Sweden | | |
| -.55 | -.55 India | | | 30% |
| | -.68 Netherlands | | | |
| -.86 | | | | 20% |

SOURCE: Husén and Mattsson, 1978.

**FIGURE 1** Standardized mean score on the attitude scale: "Belief in Science."

measuring attitudes and values:

1. *Interest in science at school.* To what extent is science liked more than other subjects and science lessons regarded as particularly stimulating and interesting?

2. *Ease of learning science.* To what extent does the student find science easy to learn?

3. *Career interest in science.* To what extent does the student aspire to a career in science in order to use science learned at school? To what extent does he or she consider science a field for creative people? To what extent does it provide good and prestigious jobs?

4. *Beneficial aspects of science.* To what extent is science considered important for social and economic development and for making the world a better place to live in? To what extent is it worth spending public money on research in science?

5. *Non-harmful aspects of science.* To what extent does the student consider science as non-harmful to environment? To what extent does science contribute to the problems besetting the world and to the creation of anxiety? Can scientific discoveries be considered to do more good than harm?

The country profiles for the three target populations (10-, 14-, and 18- to 20-year-olds) for four of the scales (ease of learning science is left out because of doubtful validity) are presented in Figure 2 on a standardized scale ranging from -100 to +100. The value of zero represents a neutral level of attitude.

As can be seen, attitudes toward science are, on the whole, positive, indicating strong support for science and the study of science. Fourteen-year-olds tend to be more positive than those in the final grades of secondary school. There is also, on the whole, a consistency in attitudes across grade levels. Students in Hungary, Italy, and Thailand tend to hold more favorable attitudes, whereas those in the Netherlands, Japan, and Sweden tend to be less favorable. These findings, along with those from other surveys, show that countries in the beginning phase of industrial and technological development generally encounter rather favorable attitudes toward science from students.

Attitudes are less favorable in countries experiencing a high level of industrial and technological development. In these countries, young people have had ample opportunity to experience the ecological effects of an advanced society. This explains, for instance, why Swedish students, in spite of relatively favorable attitudes toward science as a subject field, hold rather strong reservations about the beneficial impact of science in society and do not play down its harmful effects. Similar tendencies can be observed in the United States.

There is a trend of declining interest in *careers* in science from the 14-year-old level to the final secondary school grades in Australia, England, Finland, and Hungary. This is partly due to the retention rate and the selection that takes place in secondary schools. In Finland and Hungary, a very high proportion of girls are retained in school and their less favorable attitude toward science lowers the overall average. The polarization of students in England and Australia into science and non-science-oriented courses during the last few grades explains the drop of career interest in these countries. The relatively low interest of Japanese students at the secondary level in science careers may seem puzzling; however, one important explanation could be that the proportion of an age group entering advanced science programs at the tertiary level is by far much lower in Japan than in all the other highly industrialized countries. In Japan, the majority of students go on to technological studies.

The tricky problem of if and to what extent attitudes influence achievements in science, and if and to what extent it is the other way around has been treated with the multi-variate analysis methods that modern statistics has made available in the social sciences. The analytic approach employed allowed estimations of the reciprocal effects. By repeating the procedure in all countries and population levels, outcomes gain strength of validity.

The analyses gave the following outcomes. The effects of interest in school science are greater than the effects of achievement on interest. This has to be interpreted in the context of the problem of the quality of teaching. A major concern in science education has to do with the difficulty of recruiting competent teachers in science. In countries, such as the United States and Sweden, a high proportion of science teachers at the secondary level lack adequate subject matter preparation. Low quality teaching often may be the result, which, in turn, may affect not only achieve-

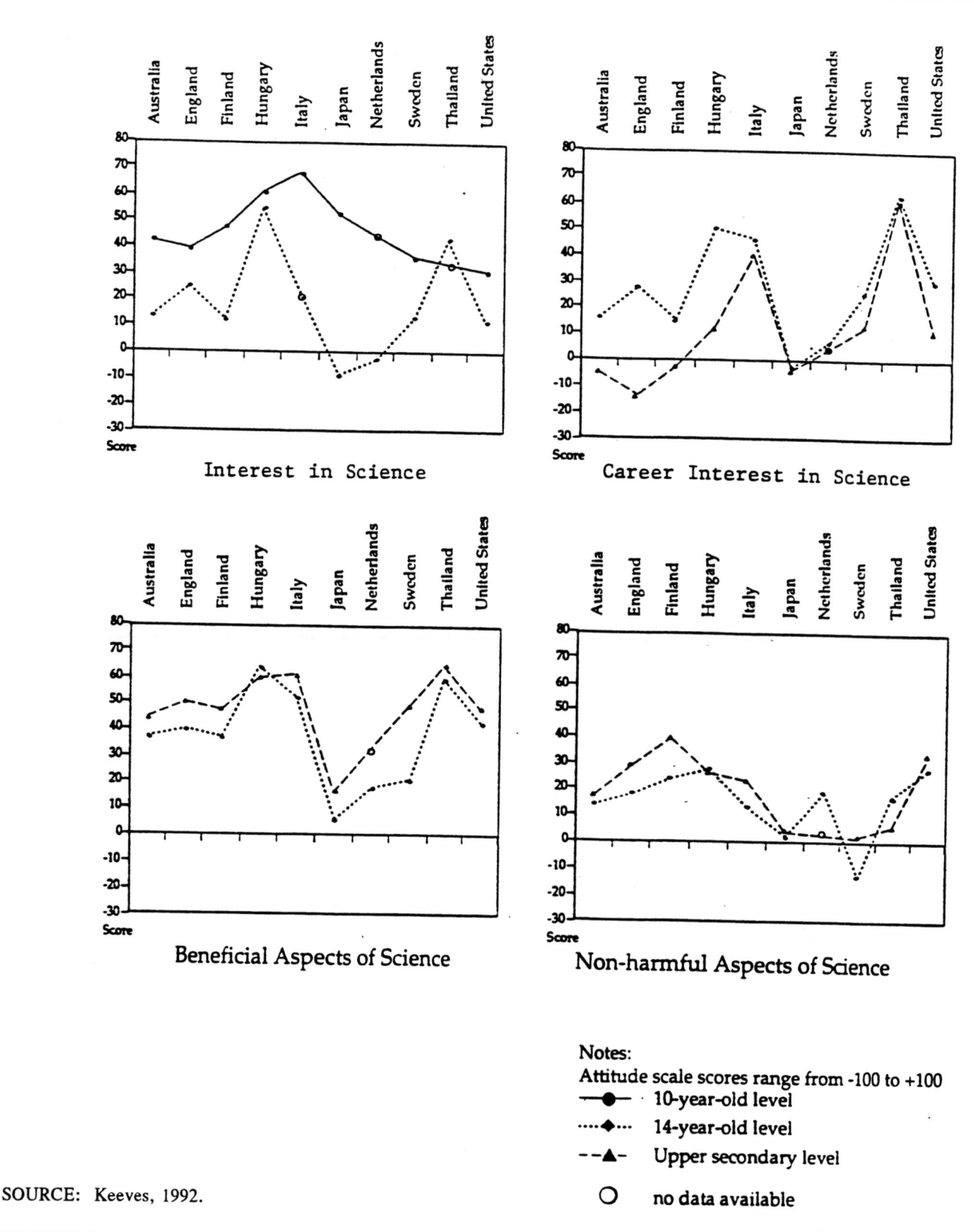

**FIGURE 2** Attitudes compared across countries and populations, 1983-1984.

SOURCE: Keeves, 1992.

Notes:
Attitude scale scores range from -100 to +100
- ●— 10-year-old level
- ····◆···· 14-year-old level
- --▲-- Upper secondary level
- ○ no data available

ments but also student interest in and attitudes toward science.

The effects of attitudes dealing with the beneficial aspects of science on achievements are of the same strength as the effects of achievements on the attitudes to the beneficial aspects.

The effects of the perception that science is easy to learn based on science achievements are significantly greater than the effects of these attitudes on achievement.

So far we have been dealing with two categories of attitudes:

1. *attitudes toward science as a school subject* and the ease, or difficulty, to learn science; and

2. *science values* as expressed in terms of the attitudes of beneficial and harmful aspects of science and career interest in science.

These attitudes and values operate at two levels, both *within* schools and classrooms and *between* schools and classrooms. Analyses were conducted at the 10-year and 14-year level, where 100 percent of the age groups is in school. Both clusters, attitudes and values, turned out to have a direct and/or indirect effect on differences in science achievements in all countries. With the exception of England and Australia, the average level of attitude toward school science and the average level of values significantly accounted for the difference between classrooms and schools.

The latter finding has important implications for the teaching of science. By keeping the students' home background and ability under control, science attitudes and values held by the student are significantly influencing the achievements in science. This attribute is an important role for the teacher in influencing attitudes. Thus, teachers can directly and indirectly influence the achievements of their students in science via the classroom climate by inspiring interest and convincing their students that science is not necessarily difficult to learn nor necessarily harmful and that it is worth investing in a science-based career.

## CAREER IMPLICATIONS OF SEX DIFFERENCES IN SCIENCE ACHIEVEMENTS AND ATTITUDES TO SCIENCE (IEA SURVEYS)

The sex differences in secondary school science participation, science achievements, and attitudes were studied in-depth by both of the international surveys on science education in 1970 (Combes and Keeves, 1973; Kelly, 1978) and in 1983-1984 (Rosier and Keeves, 1991; Postlethwaite and Wiley, 1991; and Keeves, ed., 1992). As can be seen in Figure 3, there were substantial sex difference in achievements on both occasions. The differences were, as was shown by Kelly (1978), consistent across countries and social strata within countries. The so-called effect size, which made it possible to compare age levels and countries by means of the same scale, showed the following:

1. Differences are comparatively small at the 10-year-old level. However, they increase and are rather large at the 18-year-old level.

2. Differences are relatively small in biology, intermediary in chemistry, and rather large in physics.

Given the consistency of these findings across cultures, countries, and social strata, one is tempted to arrive at the conclusion that they are genetically determined. But such a conclusion would be premature for the following reasons. First, since this is the easiest factor to investigate, we have to consider the proportions of boys and girls studying science beyond the level at which it is a mandatory subject. But taking this into account, we still find large sex differences in achievement.

At the stage when taking science is no longer mandatory, we can note large sex differences in participation in the three main subfields: biology, chemistry, and physics. Figure 4 presents the enrollment rates (for boys) in biology, chemistry, and physics in nine countries at the pre-university level. The rank order between the three subfields is consistent across all countries: the highest enrollment is in physics, chemistry is in the middle, and the lowest participation is in biology. For girls at the pre-university level, the rank order is the reverse—the same trend as achievement.

Second, we notice that there are significant reductions in sex differences over the 14-year interval between the two surveys. Countries with a significant decline at the upper secondary level are Australia, Finland, Hungary, Japan, and Sweden. In Australia and Sweden, the drop in differences at this level are particularly striking in physics. This should be looked at as the outcome of attempts to involve girls in science

CAREERS IN SCIENCE AND TECHNOLOGY: AN INTERNATIONAL PERSPECTIVE

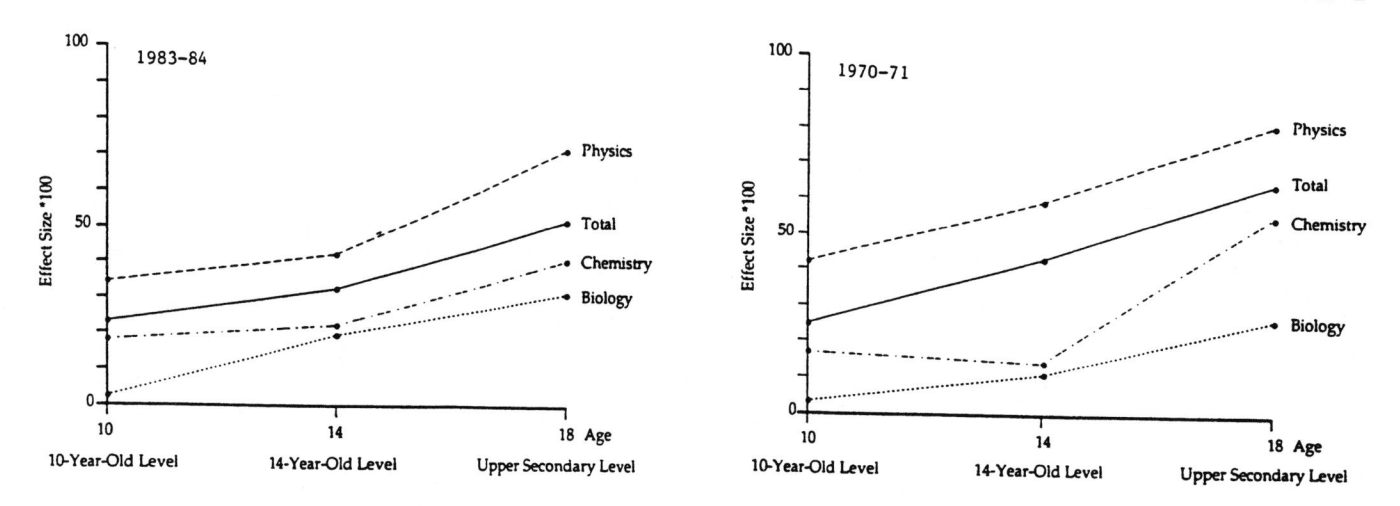

Note: The difference recorded is called an *effect size* and is obtained from:

$$\text{Effect size} = \frac{(\text{boys score} - \text{girls score}) \times 100}{\text{pooled standard deviation across countries}}$$

SOURCE: Keeves, 1992.

**FIGURE 3** Sex differences in science scores and subscores at different ages, 1970-1971 and 1983-1984.

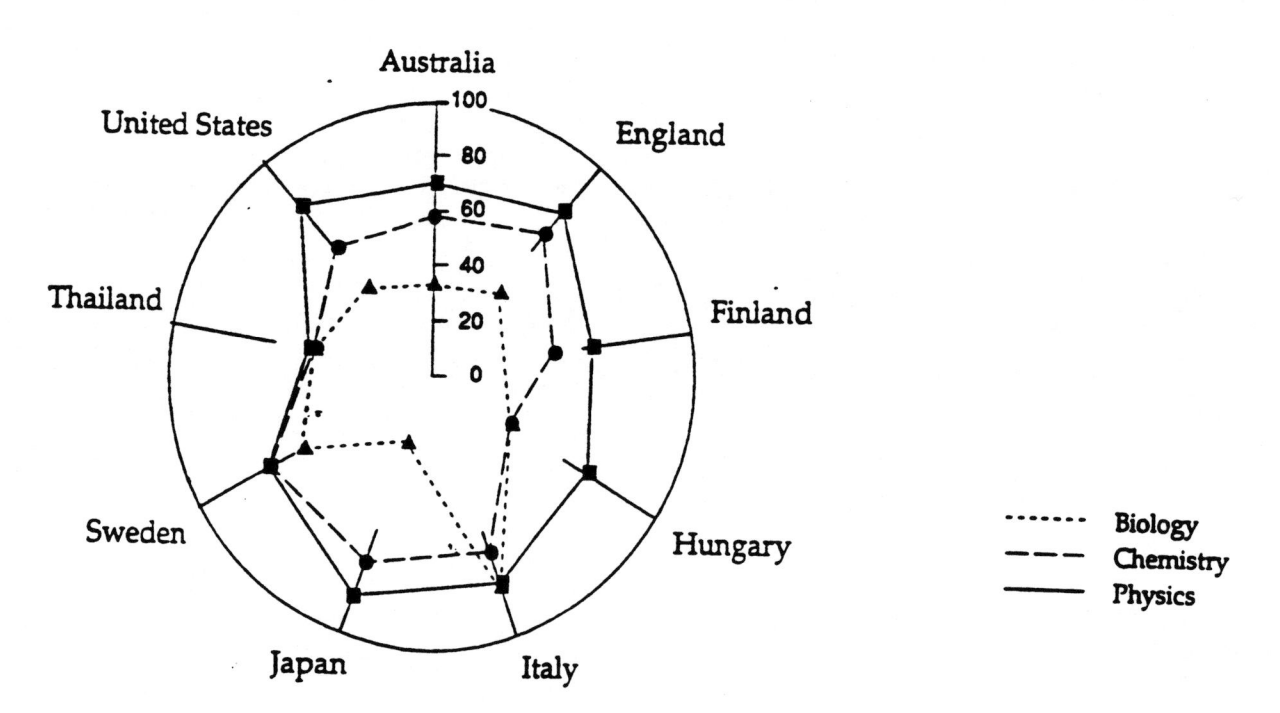

SOURCE: Keeves, 1992.

**FIGURE 4** Percentage males of total group of science specialist students studying biology, chemistry, and physics at the pre-university level, 1983-1984.

and technology programs that were launched in the 1970s in these countries.

In conjunction with the enrollment explosion taking place at the upper secondary level, the proportion of boys and girls in academic programs (including science) has changed. In some countries, boys tend to go to vocational programs that lead directly to employment, whereas girls more often take academic programs. The ratio of male to female students in academic programs in the upper secondary level in Finland and Hungary, for instance, decreased from 0.8 to 0.6 and from 0.8 to 0.6, respectively.

It was found that sex differences in attitudes toward science increase with age, a phenomenon parallel to that in achievement.

The multi-variate, causal analyses conducted show that the sex of the student has a *direct* influence on both achievements and attitudes. Thus, as students move from primary school to lower secondary and pre-university upper secondary school, a gender gap, particularly in physics, emerges. But it needs to be underlined that in the IEA surveys sex of the student has only a weak—and indirect—influence on science achievement.

## ATTITUDES AND MOTIVATIONAL ORIENTATION (NON-IEA STUDIES)

In two articles in *The New York Times* (January 25 and 26, 1993), Shirley M. Tilghman cites statistics on female scientists in the United States. In 1966, 23 percent of bachelor's degrees in science were held by women, and by 1988, this had risen to 40 percent. Women tend to choose biology instead of physics or chemistry. Thus, in 1988, 50 percent of the biology majors holding a bachelor's degree were women.

There was also an increase of women studying at the advanced level. Thus, 9 percent of the doctorates in science in 1966 were held by women, a proportion that rose to 27 percent for 1988. However, half of the increase was in psychology degrees. Little progress had been made at the graduate level in mathematics, physics, and engineering. Only 7 percent of doctorates in engineering were held by women by 1988.

Fortunately, available empirical studies, 1965-1981, with a quantitative approach dealing with correlations between gender, ability, attitudes, motivation, and achievements have been reported by Steinkamp and Maehr (1983 and 1984). They identified 66 studies published during the period 1965 to 1981, and 255 gender correlations between achievement, attitude (effect), and ability were subjected to a meta-analysis.

The gender-achievement correlations have consistently showed a small, but significant, superiority for males. Kelly (1978), in her study of sex differences in 19 countries, found the same in all countries. Interestingly, the highest differences were found in physics. In a study using a semantic differential scale, Weinreich (1977) showed that students "perceive physics, mathematics, and engineering as masculine" (Steinkamp and Maehr, 1983). The masculine image of physics is reinforced by the school where physics teachers and students are mostly male. Several studies indicated that physics is influenced more by out-of-school learning than any other branch of science. Boys are more active with appliances and engines, while girls are more interested in plants and pets.

Boys tend to be slightly superior to girls in quantitative and spatial-visual ability. In particular, there are more boys than girls among the top 5 percent of students in these abilities. Cognitive ability was positively related to achievement in science ($r=.34$) but almost unrelated to science-related attitudes, which were significantly correlated ($r=.19$) with achievement.

Steinkamp and Maehr (1983) concluded, "It all seems simple enough: one should like what one does well and do well what one likes. Simple it may be; correct it is not." The picture is much more complex when it comes to school science. Boys no doubt score slightly better than girls in both science and science-related abilities. But girls do not like science in school any less than boys. It is "primarily the acquisition of proficiency that leads to positive attitudes." Cultural stereotypes, such as science is not for girls, and expectations operate as instruments of cognitive socialization.

Regardless of whether they like science or not, girls may heed achievement goals comparable with those among boys. Thus, Steinkamp and Maehr (1984) synthesized the research literature on *motivational* orientations toward achievement in school science, aware of the disturbing, undisputed fact that women are heavily underrepresented in the scientific community or in professions based on science. Are gender differences determined by experiences at the very early school age? Can they be ascribed to motivation shaped at this stage or even earlier? Again, the existing literature between 1965 and 1981 was scanned. The search yielded 83 studies. Boys tend to be slightly superior to girls in motivational orientation, but the difference is so small that it cannot serve as the main explanation for female underrepresentation in science professions. Various dimensions of motivational orientation were elucidated by the reviewed investigations. When asked, girls responded affirmatively more often than boys that science is not just for boys. But when asked about their relationship to science, girls responded more negatively. Furthermore, girls are less frequently involved in extracurricular activities than boys. Girls think that science-related occupations are more difficult to combine with family duties. Females who would otherwise have chosen science careers are afraid of hostile male colleagues.

Thus, girls verbally object to traditional stereotypes about their relationship to science, but when they are faced with situations in which they personally have to make a choice, such as engaging in science-related achievements or embarking on careers in science, they tend to behave traditionally. It should also be pointed out that there are few female role models in science.

There are, however, as we have seen above, differences between the various branches of science with regard to motivational orientation. Girls have a stronger motivational orientation than boys in biology, whereas the opposite is the case in physics and general science. Enrollment statistics in upper secondary and tertiary education show that girls enroll more often than boys in courses dealing with life processes. Combes and Keeves (1973) had data for 10 countries showing that the proportion of women in physical science courses was much lower than that in biological science courses. One out of five doctorates in biological sciences were awarded to women but only one out of twenty in the physical sciences. The same tendency is also prevalent among academically exceptional students. Stanley found in his sample of high-ability youth that more females planned to major in biology than males, but the opposite was true in physics and engineering.

Variations across countries indirectly support a cultural or social-psychological explanation as against a genetic one for gender differences in motivational orientation toward science. The largest sex differences were found in Japan, which also had the lowest enrollment in science at the tertiary level of all the countries in the first IEA survey of girls. On the whole, the largest differences in motivational orientation were found in technologically developed countries, such as Japan, the United States, and Sweden. It is interesting to combine this finding with that by cultural anthropologists that femaleness is related to science achievements in low-achievement-motivated cultures as compared to high-achievement-motivated ones. This is contrary to generally held beliefs.

It has been pointed out that sex differences have decreased significantly in several countries from 1970-1971 to 1983-1984, particularly in countries where special efforts have been made to stimulate the participation of girls in science programs. This refutes the hypothesis close at hand that sex differences in achievement are mainly genetically determined. This brings me to trot on the thin ice of speculation, which I so far cannot support by any empirical evidence.

It is of interest to take an epistemic look at how science, particularly physics, goes about investigating the laws and secrets of nature. To what extent is the approach so far dominating in Western science a male one? Why do girls perform relatively better in biology and why do they hold relatively more favorable attitudes of this subject? Is the mode of inquiry more feminine? In most universities, girls enroll in art courses much more frequently than boys. One can account for this by pointing to the less abstract and rational way of knowing when it comes to understanding and appreciating art. The philosopher George Henrik von Wright (1983), quoted previously, pointed out that there are post-modern signs not only in art but also in other fields of inquiry. The Western dominated culture—including its scientific paradigms—has been put to question. This tendency is inspired by the loss of prestige of science due to the misuse of technology as "a consequent weakening of the intellectual curiosity which is the psychological motive for the epistemic orientation of science."

I would not have brought up his speculative view if there had not been some reason to ask ourselves whether the receding interest of young people in pursuing advanced careers in science has not had some of its inspiration from irrational sources of the kind hinted above.

All the way from Sputnik in 1957 until now, achievements in science and its applications have been seen as sharpening a nation's competitive edge in the world market. Scientific discoveries and their use have been perceived as instruments not only in improving the progress of national economies but also in establishing better conditions for individual human beings, at the very least, by improving the standard of living. We have taken this for granted recently when our globe has been beset by ecological problems that threaten the quality of life. I think that the attitude toward science and the willingness—or reluctance—of young people to embark upon careers in science ought to be examined in this context. The yielding interest in scientific careers may be an outcome of an ongoing silent revolution.

## CONCLUDING OBSERVATIONS

Factors influencing science achievement, attitudinal and motivational orientation toward science, and, in the long run, propensity to embark upon careers in science are many and operate in an intricate interplay with each other. Ability plays a role but not the most decisive one. Attitudes and motivation anchored in a particular culture are often more important. Thus, the degree of achievement-orientation, the existence of role models, and the concept of science as a difficult subject has to be considered in this context. The role of gender has been increasingly the focus of studies conducted in a field with great sex differences, particularly in physical sciences, with regard to actual achievement, motivation, enrollment patterns, and career choice.

The implications for educational and scientific policy of the research reported here are not straightforward and easy to bring out. On the one hand, it is evident that steps have to be taken to shape and stimulate motivation to study science and pursue careers in science. This relates particularly to females who are hindered by cultural stereotypes. On the other hand, there is an epistemic problem stemming from the way knowledge of science is acquired as compared to that in the humanities and arts. The two main areas of human knowledge and insight have different grammars. The way they are learned early in life, both inside and outside of school, determines how many young people will devote their life to careers in the respective fields.

## REFERENCES

Campbell-Ricardo, R. 1985. Women and Comparable Worth. Stanford, CA: Hoover Institution and Stanford University.

Combes, L. C. and J. P. Keeves. 1973. Science Achievement in Nineteen Countries. Stockholm: Almquist and Wiksell. New York: Wiley.

Engström, Jan Åke. 1994. Science Achivement and Student Interest: Determinants of Success in Science Among Swedish Complusory School Students. Stockholm University: Institute of International Education. (Studies in Comparative and International Education, No. 28).

Engström, J. A. and R. Noonan. 1990. Science Achievement and Attitudes in Swedish Schools, Studies in Educational Evaluation, 16:443-456.

Foshay, A. W. ed. 1962. Educational Achievements of Thirteen-year-olds in Twelve Countries. Hamburg. UNESCO Institute for Education.

Garfield, E. 1993. Women in Science. Part 1: The Productivity Puzzle. Current Contents, 25(9):3-5

Grasz, B. J. 1991. Report on Women in the Sciences at Harvard. Part I: Junior Faculty and Graduate Students. February 13, 1993. (mimeo)

Hurd, P. 1991. Why We Must Transform Science Education. Educational Leadership, 49(92):33-35.

Husén, T. and I. Mattsson. 1978. Ungdomars attityder till naturvetenskapen: en internationell jämförelse (Young people's attitudes towards science: An international comparison). Pp. 67-82 in P. Sörbom (ed) Attityder till tekniken (Attitudes towards technology). Stockholm: Bank of Sweden Foundation and the Royal Swedish Academy of Engineering.

Husén, T. and J. P. Keeves (eds). 1991. Issues in Science Education: Science Competence in a Social and ecological Context. Oxford: Pergamon.

Husén, T., et al. 1974. Sex Differences in Science Achievement and Attitudes. Comparative Education Review 18(2):292-304.

Husén, T. and I. Mattsson. 1978. Ungdomars attityder till naturvetenskapen. En internationell jämförelse. (Young Peoples' Attitudes Toward Science: An International Comparison). Stockholm: Liber/Allmänna förlaget.

IEA. 1991. The Third International Mathematics and Science Study. Vancouver, B.C., Faculty of Education, University of British Columbia.

IEA. 1988. Science Achievement in Seventeen Countries: A Preliminary Report. Oxford and New York: Pergamon Press.

Keeves, J. P. 1992. Learning Science in A Changing World: Cross-National Studies of Science Achievement 1970 to 1984. The Hague: The International Association for the Evaluation of Educational Achievement (IEA).

Keeves, J. P. ed. 1992. The IEA Study in Science III: Changes in Science Education and Achievement, 1970 to 1984. Oxford: Pergamon.

Kelly, A. 1978. Girls and Science: An International Study of Sex Differences in School Science Achievements. Stockholm: Almquist & Wiksell International.

Korballa, T. R. 1988. Attitude and Related concepts in Science Education. Science Education, 72(2):115-126.

Krynowsky, B. A. 1988. Problems in Assessing Student Attitude in Science Education. Science Education, 72(4):575-584.

Mählck, L. 1980. Choice of Post-Secondary Studies in a Stratified System of Education: A Swedish Follow-Up Study. Stockholm: Almqvist & Wiksell International.

McKnight, C. C. et al. 1987. The Underachieving Curriculum: Assessing U.S. School Mathematics from an International Perspective. Champaign: Stipes Publishing Co.

Oliver, J. S. and R. D. Simpson. 1988. Influences of Attitude Toward Science, Achievement Motivation, and Science Self Concept on Achievement in Science: A Longitudinal Study. Science Education, 72(2):143-155.

Postlethwaite, T. N. and D. E. Wiley. 1992. The IEA Study of Science II: Science Achievement in Twenty-Three Countries. Oxford: Pergamon Press.

Postlethwaite, T. N. and D. E. Wiley. 1991. The IEA Study in Science II: Science Achievement in Twenty-three Countries. Oxford: Pergamon.

Rennie, L. J. and K. F. Punch. 1991. The Relationship Between Affect and Achievement in Science. Journal of Research in Science Teaching, 28(2):193-209.

Rosier, M. J. and J. P. Keeves. 1991. The IEA Study in Science I: Science Education and Curricula in Twenty-three Countries. Oxford: Pergamon.

Steinkamp, M. W. and M. L. Maehr. 1984. Gender Differences in Motivational Orientations Toward Achievement in School Science: A Quantitative Synthesis. American Educational Research Journal, 21(1):39-59.

Steinkamp, M. W. and M. L. Maehr. 1983. Affect, Ability and Science Achievement: A Quantitative Synthesis of Correlational Research. Review of Educational Research. 53(3):369-396.

Steinkamp, M. W. and M. L. Maehr (eds). 1984. Women in Science. Greenwich, Conn.: JAI Press.

Tilghman, S. M. 1993. Science vs. the Female Scientist. The New York Times, January 25, 1993, p. A17.

Tilghman, S. M. 1993. Science vs. Women—A Radical Solution. The New York Times. January 26, 1993, p. A23.

Von Wright, G. H. 1989. Science, Reason and Value. Jubilee Lecture of the Royal Swedish Academy of Sciences. Document No. 49. Stockholm 1989.

Zhao, S. 1993. Chinese Scince Education: A Comparative Study of Achievements in Secondary Schools Related to Student, Home, and School Factors. Stockholm University: Institute of International Education (Studies in Comparative and International Education, No. 26).

Zuckerman, H. 1991. The Careers of Men and Women in Science. In H. Zuckerman et al. (eds) The Outer Circle: Women in the Scientific Community. New York: Norton.

# Critique of Technical Papers

## Alfred McLaren

As the discussant for Panel 3 at the Trends in Science and Technology Careers Conference, I have been tasked with providing a critique of two excellent papers submitted by Torsten Husén of the University of Stockholm and Thomas G. Whiston of the University of Sussex. I am basically a scientist by profession and have recently become concerned with the promotion of interest in science and technology (S&T) careers. I am, therefore, grateful to Mary L. Durland of Cornell University, Ithaca, New York, for her observations and contributions in the preparation of this critique. It is interesting to note that Ms. Durland was refused entry—even though she holds two Master's degrees and an all but completed dissertation—into the Ph.D. program in the Department of Science and Technology at Cornell University because "the Department was not equipped to handle a mature, interdisciplinary candidate."

Comments on Husén's and Whiston's papers will be followed by a general discussion that addresses two questions:

1. In what ways can the research described by the speakers assist us in monitoring S&T careers?

2. What further work would be needed to permit application of these research efforts to science and career studies?

### Factors Behind Choice of Advanced Studies and Careers in Science and Technology: A Synthesis of Research in Science Education
by Torsten Husén, University of Stockholm

The International Association for the Evaluation of Educational Achievement (IEA) cross-national surveys in mathematics and science, referred to in Husén's paper, provide a unique and invaluable source of data on young people's attitudes toward science. With his intimate acquaintance of these surveys and his thorough review of the research bearing on science attitudes and factors influencing participation in science in school and afterward, Husén is unquestionably authoritative. Rather than attempt to comment on the numerous observations he makes regarding student interest and achievement in science, a blanket statement might suffice: the sociocultural environment does indeed make a difference, whether it be the national environment or family and school. The fine points of direct and indirect effects of causal and correlative relationships are best left to the careful treatment they receive in the paper.

This discussion will follow Husén's lead and focus on two matters with which his is concerned: gender differences in science; and the relationship of S&T to the present ecological mess, and the effect of these on students' attitudes toward S&T careers. The former is much studied, the latter, less so.

Basically, the conclusions of gender differences in science are these:

1. Gender differences exist in attitudes, in achievement, and in entrance into S&T careers.

2. Gender differences increase with age.

3. Special programs and initiatives to encourage female participation in science do help to close the gap.

4. Gender differences can be almost entirely attributed to sociocultural differences rather than to innate biological differences.

It is clear that even in this day and age of supposed sexual equality there persist attitudes and practices that are not to women's advantage as they relate to science, first as students and then as possible career participants. Part of this may be the persistence of traditional attitudes and values despite educational and mass media efforts to the contrary. An interesting example of deep-seated habits of treating the sexes differently comes, ironically enough, in a 50-year history of the Westinghouse Science Talent Search (STS), published three years ago. Westinghouse and Science Service, the partners in this prestigious competition, have been leaders among for profit and non-profit corporations in bringing underrepresented students, including females, into science, and a very respectable percentage of its STS winners are young women. Yet, in the chapter devoted to them, they are not "young women" but "girls." The male winners are, of course, "young men" not "boys," and they (as central actors) "have parents who . . ." while the young women winners are "the daughters of . . ." therein linguistically suppressing their independent actor status. It is the ubiquity of differentiations such as this that may adversely influence a young female interested and able in a scientific field. It would appear that a considerable body of research supports this. If a young woman is not discouraged from embarking on a scientific career, she may later find herself at a disadvantage when it comes time for promotions and raises. For although she may in fact be the prime breadwinner for her family (if she has one—singleness should not be a necessity for S&T career pursuits), her participation in the science work world may be regarded as a hobby/interest matter, not the serious job/livelihood undertaking it truly is. At the institutional level, many measures have been taken to reduce such blatant discrimination. Discrimination as it now exists at the interpersonal level and is fed by personal attitudes and assumptions about the sexes, however, can only be diminished by conscious change in individual behavior.

Then there is the vicious second shift where, although her job may be as time- and energy-consuming as her mate's, the woman nevertheless returns home to a second round of work: homemaking and child care. While the man may assist, in the vast majority of cases, the woman is still the prime parent and homemaker, and spends a disproportionate amount of time on these tasks. This apparently is a worldwide phenomenon and often requires "superwoman" energy. Some couples do come to an equitable division of labor, but these are the exceptions rather than the rule. Thus, Husén's statement that "girls think that science-related occupations are more difficult to combine with family duties" is a correct assessment and a sound reason for caution in career choice.

Of interest are the large sex differences Husén discusses with regard to participation in the three main subfields: biology, chemistry, and physics (see Figure 4 in Husén's paper). For example, males favor physics and females consistently favor the biological sciences; in fact, they enroll in them increasingly from the pre-university level to the Ph.D. level. This is true in all the surveyed countries. Why this occurs is open to speculation and includes observations of young women showing more interest in the life sciences, the mode of inquiry in biology being more feminine and not perceived as masculine, as physics is. Precisely why biology is more appealing to females than physics or chemistry could stand further research. It might result in some insights about differential functioning and values between the sexes.

Monitored too should be the salaries in the subfields, with particular attention paid to any trend toward lower, non-competitive compensation in biology. Will the biological sciences show, as other occupations have, the devaluation of that endeavor because women enter it in sizable numbers?

A last point regarding gender differences in science is a sticky one. Do females in fact have differently constituted brains than men? We know that socialization rapidly sees that the brains of girls and boys receive different inputs, and they are requested to respond in different ways. Whether through birth or through socialization, the evidence does seem to indicate that the organ of thought does, with some consistency and across cultures, vary. Females tend to be right-brained in their mental functioning; men, left. The right brain, responsible for intuitive and contextual perception, for global and integrative thinking, and for certain types of verbal and artistic facility, is, according to popular scientific culture, much undervalued in today's society. So the question becomes are women undervalued because they tend to be right-brained, or is right-brainedness undervalued because it is a characteristic of women?

Modern science does appear to proceed in a way that favors and rewards left-brain activity. This is not to say that there is no place in science for right-brain functioning. In fact, the processes characteristic of the right brain produce breakthroughs when the linear, methodical left brain is stymied, and its inclination to see things in wholes rather than as isolated, fragmented parts is an ability much to be valued, especially in a mature science and in a world that is in sore need of wholeness. Perhaps women (and all right-brained people) would find science more comfortable and vice-versa if, in teaching and learning, in research, and in the communications of science, the monopoly of left-brain activity was broken.

Finally, the query "To what extent is science considered important for social and economic development and for making the world a better place to live in?" is a timely one, as is asking to what extent students consider science to blame for environmental deterioration and other social problems. That students from developed, highly industrialized societies are less enamored with science's capacity to benefit society and more inclined to pick up on its harmful effects than students from developing countries is not surprising. As Husén states, they have had ample opportunity to experience the ecological effects of advanced industry. Neither would it be surprising if they shied away from S&T careers, believing that their participation therein was hastening the world in directions they would prefer it did not go.

The best antidote to this may be to rapidly change our sciences and technologies to environmentally-considerate ones—not because we want to attract young people to S&T careers, but because the world situation cries out for it. In responding to the need for environmentally kind and restorative technologies, many new jobs will be created, and new directions in science will be stimulated. We are, in our actions if not our attitudes, extremely slow and casual about reorienting our activities. Given the magnitude and immediacy of the problem, it would be desirable if no scientist felt they could proceed without considering the possibilities for a better world inherent in their undertakings. A transformed S&T will better attract enthusiastic youth.

## Science and Technology Careers: Individual and Societal Factors Determining Choice
by Thomas Whiston, University of Sussex

Whiston is thorough in his inclusion of a considerable amount of data from a number of studies in his discussion of individual and societal factors in S&T career choice. Perhaps it is that the subject is too broad and/or that the available research is actually not focused sufficiently—at any rate, one is left with the feeling that very little can be said conclusively about career choices, save that many things affect them, which Whiston does say.

In this apparently very complex matter, it might be of help to focus first on *individual* dynamics affecting career choice, then separately on societal dynamics.

In regard to individual factors, surveys that elicit preferences, values, and intentions in a simple kind of rank order (which most of them do) produce percentages in a number of categories. In analyzing these, it is difficult to determine how the individual would actually incorporate them into career choice decisionmaking. While various kinds of statistical processing may seek to factor and weigh survey responses, these, may it be proposed, are a poor second to eliciting responses, via the surveying instrument, that give insight into how each individual actually uses his or her perceptions and preferences to come to a choice.

Eliciting responses around four main questions, and seeing these in relation to each other, might improve our understanding of how individuals perceive themselves in regard to S&T, and careers therein. The four main questions are as follows:

1. Do you find science interesting?

2. Do you feel you have the ability to go on in S&T studies? To pursue an S&T career?

3. Are you intending to go on to further education in a S&T field?

4. Is an S&T career a possibility for you?

With each of these four main questions could come a number of probes to get at the why and why not behind the stated perception or intention. These would yield a great deal of information on how the respondents have experienced science thus far in their lives and how they perceive the nature of, and opportunities in, S&T careers.

Such a questionnaire would not only reduce the second guessing that occurs when surveys do not contain within their own information-gathering parameters the responses necessary to make sense of the responses they do get, but also stimulate reflection on the part of the respondents. To probe into the whys and why nots of a given response is to come to know the factors and experiences that figure into the answer. With that information, we can better assess how societal factors, including education, are influencing S&T orientation.

Whiston provides his own suggestions for further research under his section entitled "Future Research Tasks." It is not that these would not be worthy undertakings, but whether they would get to the crux of the matter regarding individual choice as well as the four straightforward questions is questionable.

While there is much to comment on in a paper the scope of Whiston's, time does not allow. Further commentary is in fact incorporated under the general discussion section of this paper. Before moving to a consideration of societal factors in S&T choice, one further observation begs inclusion. Percentage-wise, trends show that young people tend to choose S&T less and fields such as social science and communications more. Perhaps we should keep in mind, when viewing this trend, that while the world of S&T is largely occupied by those producing such items, increasingly there is a need for, and are, science-affiliated occupations, such as management, communications, etc., which are in fact essential parts of S&T in today's world. Thus, the education and orientation of young people toward science so that they may be science-informed, if not science-productive, is of great importance. Attention to science-affiliated careers should be considered part of the task within the betterment of S&T careers overall.

In regard to societal factors including choice, Whiston puts forth many variables. He also details a number of "Policies to Improve S&T Literacy (and Possible Selection of S&T as a Career)" in his paper. No doubt solid, rewarding teaching; better ties between industry and education; coherent national policy; and the like would all be of benefit.

There may be an overarching circumstance that must be dealt with in order for these policies to have their desired effect. That circumstance is the contraction of the economies of many nations, along with prolonged recessions, and general economic bad times. We like to think skilled S&T manpower is much in demand and that more is needed. But the actuality, as experienced by those already in S&T careers and suspected by those who are considering them, is that jobs are sometimes very hard to find, and even a noteworthy previous career does not necessarily guarantee continued employment. Good S&T jobs may be especially hard to come by, with lateral and downward mobility being frequent. In this environment, it is no wonder that young people gravitate toward those occupations, and preparations for them, that they perceive (correctly or incorrectly) as offering reasonable chances of employment. A corollary of this is that it appears that some students may be opting for *no* preparation, as any preparation seems like a long shot, so why bother. This certainly appears to be true in the United States, where a substantial number of students are not sufficiently motivated to become truly literate, let alone specialize. It is highly doubtful that this is the fault of the education system.

What needs to be done is to reestablish the correlation between education and jobs, which has in truth been severely affected by recent economic events. Our youth and their talents cannot be treated as commodities in a free market economy if we wish to have them take education seriously, and this includes science education. Measures that will guarantee at least entry-level employment need to be taken; and if nations and corporations are so convinced that they need more skilled manpower, they must structure their occupational worlds to accommodate it. Until this is done, perhaps it would be prudent to regard S&T careers as subject to, along with everything else in the world, overpopulation.

## GENERAL DISCUSSION

The best way to monitor trends in S&T careers, in my opinion, is to tap the collective experience of those already in them. Panel 3, devoted as it is to factors influencing choice among young people as they do or do not select S&T careers as desirable ones in which to realize their abilities and aspirations, rightly does not have much to say about the careers themselves. It is for this conference on the Trends in Science and Technology Careers overall to do this; and from this conference, those concerned with the recruitment of competent young persons into S&T should be able to gain further insight into how best to modify and update the image of S&T careers that is presented, either intentionally or indirectly through the education they receive, to potential S&T career entrants.

In looking at career choice, there is always the question whether young men and women make their selections based on the reality of a given occupation or on other factors, such as an idealized or otherwise incorrect image, or on the supposition that what they did well in and liked in school will transpose via entrance into an occupation and ultimately into a satisfactory livelihood. What this panel session does offer is a chance to discuss whether S&T careers as they currently exist (and are likely to in the near future) do in fact offer young people what they desire. Together, the Husén and Whiston papers, composed of statistics from a number of sizable surveys, present a picture of considerations in career choice.

There is a temptation in science to feel that no matter how much data has been collected, more is needed, or at least desirable, and that the solution to a problem is to be found through further research. Indeed, this—suggestions for further studies of factors in S&T career choice—is what has been requested of the discussant. May it be proposed that what is needed is not more research, but action, based on what we know and suspect already. We already know the following:

1. The choice of a career is not made on any single factor alone; a number of things contribute to any given career choice.

2. The choice is normally not made once and for all at a given point in time; experiences and perceptions accumulate and change, and ultimately add up, or fail to, in favor of an S&T career.

3. There are many possible points of intervention in the contexts that provide experience in science and that generate attitudes toward S&T careers.

4. Intervening at these points and making changes in the experiences young people have with regard to science, particularly as they provide opportunity for a broad and realistic picture of S&T in the world, will improve the recruitment of suitable individuals to S&T careers.

5. The exception to this will be if S&T careers are in reality not desirable.

Let us start with this last item. We live in an ever-changing world, a world full of discontinuities and surprises. As this translates to occupations, and S&T careers are no exception, it means that a person cannot enter any field with the expectation that it will provide a lifetime of growth, intrinsic gratification, and external remunerations sufficient for material well-being. In the face of this reality, one that young people seem to sense, the very basic "being well off financially" has crept up through the decades to be an essential or very important factor to over 80 percent of them, while the more luxurious "develop a meaningful philosophy of life" has dropped from over 80 percent to 40 percent. Supporting this is the Boyd and Kirkland data in which "a job that gives me good long-term career opportunities" was the *number one* choice factor, with "a high future salary" being a close second. "The opportunity to be creative and original" did come in third, due largely to its importance to *non-science* majors. Thus, if S&T wants to attract young people, especially those with science backgrounds, it had better see to the stability of employment in S&T careers once entered.

Now, immediately, you who are scientists and engineers and technicians will say the following:

> The stability and continuity and salaries within our occupations are not our doing. They (or their lack) are due to factors far beyond our control, such as government priorities and funding, and shifts in research and development as the result of new knowledge and changing demands. And besides (if the truth be known), we are too busy trying to retain and stabilize our own careers to attend to such things.

All true, but this does not negate the observation that if S&T careers are to appeal to the coming generation, they must offer a degree of stability and continuity. S&T degrees are realistically perceived by some young people as being too specialized to permit extra-science career possibilities and too demanding to undertake them without a reasonable assurance of employment once achieved. While institutions and corporations worry about a qualified workforce, young people worry about long-term occupational opportunity. It may be inferred that they will shy away from careers that appear to offer entry-level positions to many, but continuing employment opportunity to only a few.

S&T organizations need to be structured so that they offer a number of positions at a number of levels, with responsibilities and rewards increasing gradually and incrementally. Too often there is a tendency for organizations to become feudal in stratification, with a few amply rewarded individuals at the top and a number of workers consigned to low-level though skilled tasks that are perceived in many cases to be dispensable and, should they demand too much, replaceable. S&T can be and is often conducted this way with little harm to the task at hand. However, considerable harm is done to manpower resources when such qualities as originality and broad experience are not encouraged. This ultimately results in great harm being done to all young people who would seek a satisfying, lifelong career, which most consider to be of paramount importance.

If scientists and technologists cannot or do not want to create suitable organizations themselves, they need to hire science-informed individuals who will; and all need to argue for S&T policy that will accomplish it.

This discussion may seem like a long detour taken at the expense of traveling down what appears to be the main highway of education and its effects on S&T career choice. However, along that highway there *are* many intervention points, and few would argue that they shouldn't be taken advantage of and that the highway itself is variously in need of repair, upgrading, re-routing, and access ramps for those who have had unequal opportunity to use it. But if S&T careers are not in fact good places for long-term, reasonably rewarded employment, then even the best science education system will not produce new entrants into S&T careers.

# PART IV

Utilizing Points of Intervention to Enhance and Sustain Interest in Science and Technology Careers

# Introduction to Utilizing Points of Intervention to Enhance and Sustain Interest in Science and Technology Careers

Pim Fenger[1]

Under this topic we are going to concentrate on the science and technology (S&T) track and the decision points along which an individual may opt in or out of the career path. We have been asked to concentrate on mechanisms that have been shown to have significant effects in increasing the likelihood of an individual moving along the track.

In order to highlight this subject, I have to confess that for governments these questions are not new at all. Most OECD countries have the rather uncomfortable feeling that they are entering a period of demand for highly qualified research personnel, especially in the scientific and engineering fields, which may exceed the available supply.

In the Netherlands, for example, warnings can be heard about structural shortages in the supply of researchers and engineers on the medium long-term (Advisory Council for Science and Technology Policy).[2] A ROA study on the labor market for research in 1990-2010 (Berendsen and Willems)[3] shows that several shortages are to be expected in the hard sciences after 1995.[4] The IRDAC study, *Skill Shortages in Europe*, indicates a low share of engineers in European industry as compared to that of the United States and Japan.[5]

Under the actual circumstances of weak economic growth, the existing labor market tensions are still acceptable. But, when economic growth starts gaining momentum again, the shortage of skilled technicians and researchers will be felt. The demographic development (in the year 2000 the count will be 25 percent less youngsters in the 18-24-year-old category), combined with the diminishing interest of students in natural sciences, shall aggravate the situation. This may have dramatic effects for the industry and the economy, in general, in the Netherlands. The Advisory Council for Science and Technology Policy in the Netherlands tells us that active recruitment of research and development (R&D) personnel from Eastern Europe, or the Far East, probably cannot be avoided.

However, not all experts share this view. Lemstra de-dramatizes the situation that can be expected.[6] As he points out, the supposed shortages are based on extrapolations of figures from the late 1980s, an economic boom period with high investment rates into new projects—the chemical sector in particular. Corporate research laboratories smoothly accommodated new generations of chemists, physicists, and engineers. Now, in 1993, the situation has changed. The opening up of Eastern European markets gave rise to cheap imports from these countries with negative effects on the growth rates of western bulk chemical and steel industry. Money for investment in research projects involving large risks is lacking. Besides, the increased environmental consciousness of society has led to political priorities favoring recycled materials. Composite and complex materials are squared to the concept of recycling. Lemstra foresees not so much a quantitative problem as qualitative deficiencies. Improvements in the quality of formation should start at the teacher level in secondary education.

Government concerns have been very well expressed in the 1989 OECD report, *Research Manpower: Managing Supply and Demand*.[7] This report was prepared by R. J. Kavanagh (Canada) and

Alan Fechter (United States) under the aegis of the OECD Committee for Scientific and Technological Policy, by its Group on Scientific and University Research. The report learned that within member countries two categories of action can be distinguished between supply and demand for research manpower: recruitment methods and retention methods.

## RECRUITMENT METHODS

Examples reported of recruitment methods include: persuading more women to choose science and engineering careers; improving the quality of science and mathematics teaching in schools; encouraging a greater proportion of young persons to follow science and mathematics courses while in school so that, subsequently, they will be qualified to enter university courses in science or engineering; encouraging more first degree graduates to continue postgraduate training; providing opportunities for employing scientists and engineers to return to university in order to undertake postgraduate studies; providing special opportunities for children from certain social, racial, or ethnic groups to overcome historical barriers and enter science or engineering education programs; and creating a greater awareness of the opportunities of challenging careers in R&D.

## RETENTION METHODS

Techniques geared to reducing the attrition of science and engineering students, and retaining research-trained persons in the R&D workforce by providing appropriate employment opportunities, can be referred to as retention methods. In general, these techniques seek to reduce the waste of potential research personnel from the educational system and train research personnel from the pool of such persons. Techniques falling into this category include methods of ensuring that as many qualified students as possible complete their science and engineering degrees at both the undergraduate and postgraduate levels, methods of preventing the exit of qualified doctoral graduates in certain fields where temporary surpluses exist from research enterprises, and policies affecting the retention of students from other countries.

From the viewpoint of the government, the added value of this panel may be that by studying some of the techniques mentioned in the OECD report in more depth and, above all, by assessing some techniques, the governments may know and understand more than they did three years ago. It is the role of our speakers, Kazuo Ishizaka and Dervilla Donnelly, to help us with this assessment in our discussion following their contributions.

Another reason why government is interested in the results of this conference originates from the Technology-Economy Program (TEP) of OECD.[8] The impressive TEP reports on technology in a changing world are focused on the relationships between science, technology, and economic growth. The TEP reports and policy recommendations (accepted by the OECD Ministerial Council in 1991) have widened the scope of R&D policies in many countries. Traditionally, S&T policies in advanced countries were conceptualized within the framework of the so-called knowledge trajectory (Dosi)[9]—a continuum with basic research at the far left, mission-oriented research and applied research in the middle, and developmental work at the far right. To optimize the utility of all these research activities and their interrelationships, S&T policies were concentrated on the transfer mechanisms between the various activities. TEP, however, has added at least three important dimensions to R&D policies when it comes to their contribution to economic growth:

1. TEP stressed the importance of social acceptance of new technologies, and the knowledge and abilities of the working to handle new technologies.

2. TEP made us fully aware of the fact that all those institutions with their own distinctive missions on the knowledge trajectory had to be staffed by people with the right qualifications.

3. Under the influence of TEP, the concept of knowledge transfer is being translated and operationalized more and more toward the concept of people transfer.[10]

As a result of this very differentiated approach to

human resources, we see that R&D policies are becoming more open to and integrated with educational polices for basic, secondary, and vocational training.

A third reason for government interest in the assessment of points of intervention lies in recent policies for research training within and in networks involving universities. The development of universities in the past two centuries is marked by two important innovations. The first was the introduction of explicit research tasks within universities: the combination of educational and research tasks within one institution. This innovation of Von Humboldt dates back to 1808, after the failure of an earlier attempt by the French encyclopedists.[11] And, since then, universities have picked up their research mission in a remarkable way. The second social innovation of tremendous importance was the introduction of the idea of mass higher education after World War II. Of course the combination of education and research, and the demand for mass higher education cannot be managed easily (Hazeu).[12]

Policy initiatives in Europe, now being taken, to let universities fulfill both missions of higher education and functionally-related educational and research tasks are strongly similar. In general terms, these developments can be considered as the introduction of the American graduate school model on a European scale. Good examples are the Graduierten Kollegs in Germany, the Ecoles Doctorales in France, the Networks in Belgium, and the Research Schools in the Netherlands.[13] Generally, they can be seen as an attempt to make Ph.D. training relevant to a wider range of occupational positions than has traditionally been the case in Europe. On the other hand, they are also inspired by a shortage of teachers in higher education, which can be expected within the next decade (Blume).[14]

In my own country, the main reasons for the emergence of researchers were the needs for (1) training top quality researchers, (2) structuring research training within the second phase, (3) improving the national research infrastructure to compete better in the international field, (4) strategic concentration of scarce research capacity to avoid fragmentation and generating a critical mass by means of cooperation between universities themselves and the related research environment, and (5) selecting to improve quality of both trainer and supervisor (Hazeu).[15]

In order to receive accreditation by a committee attached to the Royal Netherlands Academy of Arts and Science, a research school must meet 10 characteristics, among which we find a research school as a real center of excellent research, having a minimal size of 40-50 research trainees (critical-mass), and being part of one or more universities. Furthermore, a research school should be an independent organizational unit with its own budget responsibilities. In this way, a research school can meet the expectation that, where possible, it should cooperate with research organizations outside the university system (Berendsen and Willems). And, indeed, one of the criteria for accreditation is the requirement that a research school has to make clear that it has seriously considered multi-year cooperation agreements with research institutions of excellent quality outside the university system. This may lead to a major improvement in the infrastructure of research training.[16] For instance, the Philips Physical Laboratory has offered to open its facilities to research training ends.

Such types of efficient uses of national scarce resources that provide the right scientific climate that stimulates research and research training can also be found outside Europe. Good examples can be found in Australia and Canada. In Canada, for instance, networks of Centers of Excellence have been created since 1988 that link the facilities in universities, in government research establishments, and in industry (Blume). A report of these new forms of research training is now being prepared by OECD. Stuart Blume[17], Professor of Science Dynamics at the University of Amsterdam, has been appointed as the rapporteur. Hopefully the conclusions of this conference and the results and policy recommendations of the OECD report will reinforce each other, thus having further impact on the work of the OECD Group on the Science System,[18] as well as on national policies.

## NOTES

1. This contribution does not commit any authority. Thanks are due to Jacky R. Bax.

2. Advisory Council for Science and Technology Policy (AWT). 1992. Technici en Onderzoekers: kwaliteit en kwantiteit. SDU. Den Haag.

3. Berendsen, H. A. de Grip and E. J. T. A. Willems. 1991. De Arbeidsmarkt voor Onderzoekers (The labor market for researchers). 1990-2010. ROA. Den Haag.

4. Supply surplus and shortages of researchers in the hard sciences: totals and as percentage of expected employment for researchers in that category.

| Education | 1990-1995 total | % | 2006-2010 total | % |
|---|---|---|---|---|
| University | -580 | -2 | -5.250 | -11 |
| of which: | | | | |
| Agriculture | 880 | 37 | 200 | 6 |
| Natural Sciences and Mathematics | -1.570 | -14 | -3.560 | -22 |
| Technical | 750 | 6 | -960 | -5 |
| Medical | -640 | -7 | -930 | -9 |
| Higher Vocational | -2.490 | -8 | -7.320 | -16 |
| Secondary | -3.380 | -17 | -4.970 | -20 |

SOURCE: ROA (see Note 2).

5. Industrial R&D Advisory Committee of the CEC (IRDAC). 1990. Skill Shortages in Europe. Brussels.

6. Lemstra, P. J. 1992. Toekomstig tekort aan ingenieurs wordt overschat. In VSNU-opinie, Februari 1993. Utrecht.

7. OECD. 1989. Research Manpower: Managing Supply and Demand. Paris.

8. OECD. 1991. Technology-Economy Programme (TEP). Technology in a Changing World. Paris. OECD. 1991. TEP; Technology and Productivity. Paris.

9. Dosi, G. 1982. Technological Paradigms and Technological Trajectories. In: Research Policy 11; 147.

10. See for instance IRDAC (1992) Opinion on Biomedical Research. Brussels. We can also refer to working documents of the Commission (EC) on the 4th Framework programme, where the mobility of researchers is explicitly seen as a critical variable in the transfer of technology.

11. Fenger, P. 1992. Tradition on a Variable in Systems Analysis: The Case of the Universities. In: World Futures. Vol. 34. Gordon and Breach. New York.

12. Hazeu, C. A. 1991. Research Policy and the Shaping of Research Schools in the Netherlands. In: Higher Education Management, OECD. Vol. 3, No. 3. Paris.

13. Fenger, P. 1992. Graduate Research Training in a Number of European Countries and the United States. In: Bulletin de Methodologie Sociologique, nr. 34, March, Paris.

14. Blume, S. S., (rapporteur). 1991. Postgraduate Research Training Today: Emerging Structures for a Changing Europe. Report of the Temporary International Consultative Committee on New Organisational Forms of Graduate Research Training. The Hague.

15. See Note 11.

16. Fenger, P. 1993. University and Non-university Opportunities for Postgraduate Research Training. Forum Sozialforschung. Vienna.

17. Papers prepared for the OECD Workshop on Research Training (chair prof. S. S. Blume, University of Amsterdam) Amsterdam 17, 18 March 1993.

18. The OECD Committee on Science and Technology Policy decided in March 1993 to change the name of its Group on University and Research Policy to the Group on the Science System.

# Human Development of Science and Technology in Japan: From the Classroom to the Business World

Kazuo Ishizaka

## INTRODUCTION

Japan has experienced two major educational reforms. The first took place in 1872, some 120 years ago under the Meiji Restoration, establishing a modern, multi-line education system. The second major reform took place immediately after World War II, implementing the United States' "single-line 6-3-3-4 system" as a model.

Although the modern education system actually started after Japan abandoned its isolation policy under the Tokugawa Shogunate feudal system, there were prior to that well over 20,000 to 50,000 "Terakoyas," a form of very small private schools with one or two teachers, throughout the country. These Terakoyas and other forms of schools became a solid foundation for the first educational reform.

From the outset of the first reform in 1872, the Meiji Restoration government started a democratic primary education open to any child of any school attendance district regardless of sex, social status, or means.

Both of the major educational reforms, however, were accomplished in extraordinary social and political situations. The intended third reform has started from a rather different situation. Since the second reform, various improvements to education have been made. Among others, the Central Council on Education (CCE), an advisory body to the Minister of Education, Science, and Culture (MOE), studied basic educational policies and planning at the request of the Education Minister.

In 1971, CCE submitted its twenty-second report, *Basic Guidelines for the Development of Integrated Education System Suited for Contemporary Society* (frequently called *Yonroku Tosin*). This report was intended to make revolutionary improvements to Japanese education from kindergarten through university. However, Japan experienced more rapid changes than expected. The change in our society has been quite rapid, causing a number of problems to arise in various stages of education. In order to respond to the changing social circumstances, the national government decided to establish an Ad Hoc Council on Educational Reform [now officially called the National Council on Educational Reform (NCER)] in August 1984, directly under the Prime Minister's Office, to reexamine our educational policies and practices in order to make the third major educational reform in our history. The council did a comprehensive study on the various government policies in education and other related fields, and, on the basis of these studies, submitted four sets of recommendations to the Prime Minister. After submitting the final report, NCER disbanded in August 1987.

### Statistical Outline of Current Japanese Education

The present Japanese school system, brought into being during the postwar U.S. armed forces occupation between 1947 and 1950, provides for six years of primary/elementary school and three years of lower secondary school (junior high school), followed by a non-compulsory three years of upper secondary school and four years (six in the medical field) of university.

Since Japan implemented the U.S. model, a number of new types of schools were added to meet the needs of our changing society. There are two-/three-year junior colleges (in 1950), five-year technical colleges (1962) for graduates of lower secondary schools, and special training schools (1976) for graduates of both lower and upper secondary schools.

Japanese elementary schools, lower secondary schools, and technical colleges are public dominant, while kindergartens, junior colleges and universities, and special training schools are private dominant in terms of school enrollment. As of May 1991, ratios of students enrolled in private elementary schools, lower secondary schools, upper secondary schools, and technical colleges were 0.710 percent, 4.07 percent, 28.9 percent, and 5.7 percent, respectively; however, the latter enrollment shares of private institutions were 78.9 percent, 91.9 percent, 73.0 percent, and 94.5 percent, respectively.

No vocational/technical courses are offered during the first nine years of compulsory schooling. Moreover, there is no tracking system in public elementary and lower secondary schools. Since May 1991, approximately 74 percent of the students in upper secondary schools were enrolled in general courses and 13.6 percent were enrolled in technical courses.

A little over 90 percent (1,803,221) of 18-year-old graduates of upper secondary schools and about 54 percent of the upper secondary graduates advance to higher institutions, including 25.5 percent to 4-year universities and 12.2 percent to junior colleges, as of May 1991. The annual number of graduates from 4-year colleges and universities is 428,079. In May 1991, 14,217 (3.32 percent), 86,115 (20.1 percent), and 14,854 (3.47 percent) of these students were in the fields of science, technology, and agriculture (MOE, 1992).

## RECENT EFFORTS AT THE NATIONAL INSTITUTE FOR EDUCATIONAL RESEARCH

The National Institute for Educational Research (NIER) of Japan was established in 1949 as an organ of the MOE. NIER has been conducting basic and applied research on education in a wide range of educational fields. Based on the recommendation of the NCER that worked directly under the Prime Minister's Office during 1984-1987, NIER was reorganized in May 1989 and is expected to strengthen its function to provide foundations for educational policies as well as to enrich its function as a national curriculum center. Some of the concrete measures now underway are as follows:

- Conduct work and basic research on educational policies and practices.

- Conduct research on school curricula, teaching materials, teaching methods, and other educational topics for elementary, lower, and upper secondary schools so that the research results can be utilized by the nationwide education systems and schools.

- Strengthen cooperation with nationwide educational research institutions [National Federation of Educational Research Institutes (NFERI)], as well as international organizations such as UNESCO, OECD, and the International Association for the Evaluation of Educational Achievement (IEA).

As of June 1992, 267 prefectural, municipal, and private educational research institutions in Japan were affiliated with NFERI, whose headquarters is within NIER and where the Director-General of NIER serves as a chairman. NFERI has been conducting nationwide joint research projects, mostly in school education.

Since joining the IEA, NIER has participated in the following IEA studies:

- First International Mathematics Studies (1962-1967)

- First International Science Studies (1966-1971)

- Second International Mathematics Studies (1975-1981)

- Second International Science Studies (1980-1988)

- Computers in Education Study (1987- )

- Third International Mathematics and Science Study (1991- )

As one of the Associated Centers in Japan for UNESCO's Asia and the Pacific Program of Educational Innovation for Development, NIER sponsors several times a year workshops and seminars for educators from Asia and the Pacific regions. For example, a 1992 workshop was entitled "Regional Seminar on Goals, Aims, and Objectives of Secondary Education in Asia and the Pacific," and held at NIER from June 12-27, 1992.

## PRE- AND IN-SERVICE TRAINING OF SCHOOL TEACHERS: GRADES 1-12

### Current Pre-service Training: Mathematics and Science Teachers

Elementary and secondary school teachers in Japan are trained at universities and junior colleges approved by the MOE. The majority of teachers are currently trained at colleges and universities. Although teaching certificates are issued by each prefecture, they are valid for life and for every school in Japan.

Teacher training in Japan started with the establishment of a "normal school" in Tokyo in 1872 when Japan implemented the European multi-line 6-5-3-3 education system. For the next 80 years, elementary school teachers (for grades 1-6 and 7-8, the latter being upper elementary schools) were trained mainly by normal schools that were established in every prefecture in Japan. Middle school teachers (grades 7-11) were trained mainly by higher "normal schools."

The teacher training system was completely revised on the advice of the United States Education Mission immediately after World War II. Since the Educational Personnel Certification Law promulgated in 1949, teacher training has generally been carried out at junior colleges and four-year colleges and universities. The majority of current teachers have been trained at four-year colleges and universities. Normal schools were reorganized as four-year teacher colleges (Gakugei Universities), and the majority of current elementary school teachers are graduates of National Gakugei Universities.

The 1949 revised pre-service teacher training curriculum had three basic elements: general education, teacher education, and professional education. The teacher training curriculum was improved in 1990 to meet the needs of a changing society. According to a revised law, there are now two kinds of teaching certificates: a regular teaching certificate and a temporary one. The former is valid for all prefectures for life and is divided into first and second class. A temporary certificate is issued when authorities cannot find teachers holding regular certificates and is valid for only three years. However, teacher shortages are extremely rare and temporary certificates rarely issued.

*The basic requirement for the first class* certificate for kindergarten, elementary, and lower secondary schools is a bachelor's degree, while for upper secondary schools a master's degree, or the completion of non-degree courses for graduate school (Senkoka), is required.

*The basic requirement for the second class* certificate for kindergarten, elementary, and lower secondary schools is to be a graduate of a two-year junior college, while for upper secondary schools a bachelor's degree is required.

Table 1 shows the minimum requirements for a regular teaching certificate for elementary, lower, and upper secondary school teachers, and Table 2 shows the minimum requirements, courses, and credit requirements for mathematics and science teaching certificates. A lecture class of 15 hours, requiring 30 hours of student preparation, yields 1 credit. A seminar class of 30 hours, requiring 15 hours of student preparation, yields 1 credit. Laboratory classes require 45 hours in the laboratory for one credit.

Professional subjects include educational principles, educational psychology, teaching methods, studies of moral education, teaching practice (student teaching), etc.

**TABLE 1** Minimum Requirements for Regular Teacher Certificates

| | | | Credits to be Earned | |
|---|---|---|---|---|
| Types of Certificate | Basic Qualification | | Teaching Subject | Profess Subject |
| Elementary Schools (G1-6) | 1st class | Bachelor's Degree | 16 | 32 |
| | 2nd class | Associate B.A. or equivalent | 8 | 22 |
| Lower S.S. (G7-9) | 1st class | Bachelor's Degree Bachelor's Degree | 40 (A type) 32 (B type) | 14 14 |
| | 2nd class | Associate B.A. or equivalent | 20 (A type) 16 (B type) | 10 10 |
| Upper S.S. (G10-12) | 1st class | Master's Degree or equivalent | 62 (A type) 16 (B type) | 14 14 |
| | 2nd class | Bachelor's Degree Bachelor's Degree | 40 (A type) 32 (B type) | 14 14 |

NOTE: A-type certificates are for **science**, social studies, etc.
B-type certificates are for **mathematics**, Japanese language, etc.

**TABLE 2** Major Subject Studies Required for Mathematics and Science (as of April 1986)

| Subject for Certificate | Major Subject Studies | Minimum Number of Credits |
|---|---|---|
| Mathematics | Algebra | 4 |
| | Geometry | 4 |
| | Analysis | 4 |
| | Statistics | 2 |
| | Surveying | 2 |
| | TOTAL | 20 |
| Science | Physics (including exp) | 5 |
| | Chemistry (including exp) | 5 |
| | Biology (including exp) | 5 |
| | Geology (including exp) | 5 |
| | TOTAL | 20 |

## Upcoming Pre-service Training: Mathematics and Science Teachers

In order to upgrade the quality of school teachers, the Educational Personnel Certification Law was partially revised in December 1988 to introduce an "advanced kind" teaching certificate that commands a master's degree as a basic requirement. The law was implemented for freshmen entering April 1, 1990, and a complete revision of the law was made on April 2, 1991. The following article is based on the revised 1991 Educational Personnel Certification Law.

New teaching certificates may be classified into three major categories: regular, special, and temporary. The *regular certificate* is subdivided into three kinds—the advanced, the first, and the second—as shown in Table 3.

Tables 4 and 5 show the minimum curriculum requirements for teachers of mathematics, science, and other technical fields. The *special certificate* is designed to attract those who are working in the non-teaching sectors, who have an interest in the teaching profession, and who can bring knowledge and techniques into the classroom. The *temporary certificate* serves the same purpose as the previously stated Educational Personnel Certification Law.

Teachers are required to be trained in line with the special law for educational personnel, and various in-service training programs have been developed nationwide. For example, MOE (Monbusho) offers annual training courses for national and local public school principals and head teachers. Approximately

**TABLE 3** Minimum Requirements for Revised Teacher Certificates

| Types of Certificates | Basic Qualification | | Credits to be Earned | | |
|---|---|---|---|---|---|
| | | | Tchg Subj (TS) | Prof Subj (PS) | TS/PS |
| Elementary Schools (G1-6) | Advd kind | Master's Degree | 18 | 41 | 24 |
| | 1st kind | Bachelor's Degree | 18 | 41 | |
| | 2nd kind | Associate B.A. or equivalent | 10 | 27 | |
| Lower S.S. (G7-9) | Advd kind | Master's Degree | 40 | 19 | 24 |
| | 1st kind | Bachelor's Degree | 40 | 19 | |
| | 2nd kind | Associate B.A. | 20 | 15 | |
| Upper S.S. (G10-12) | Advd kind | Master's Degree | 40 | 19 | 24 |
| | 1st kind | Bachelor's Degree | 40 | 19 | |

5,000 school teachers are selected annually from all parts of Japan and sent to various countries to gain an international viewpoint.

Each prefecture and municipal board of education has its own systematic in-service program. Every prefecture has at least one large education center to provide in-service training as well as research activities.

A number of teachers can obtain scholarships to further their studies. For example, a little over 400 teachers annually obtain scholarships to work toward master's degrees. Approximately 90 percent of them study at the newly-established graduate schools designed for teachers.

Based on a law promulgated in May 1988, a new and strong in-service program has been implemented to provide all new teachers with one full year of training. During the 1990 academic year, one year of compulsory training started for all new teachers of public elementary and lower secondary schools. Various other forms of in-service training programs have been designed by national, local, or school levels to meet the changing social and technological circumstances.

## CERTIFICATION SYSTEM OF TEACHERS OF HIGHER EDUCATION

Qualifications for teachers at colleges and universities (including graduate school), junior colleges, and technical colleges are specified in the MOE ordinances entitled "Standards for the Establishment of

TABLE 4 Major Subject Studies Required for Mathematics and Science for Lower Secondary Teachers (as of April 2, 1991)

| Subject for Certificate | Major Subject Studies | Minimum Number of Credits |
|---|---|---|
| Mathematics | Algebra | 6 or 4 |
| | Geometry | 6 or 4 |
| | Analysis | 4 |
| | Probability & Statistics | 4 or 2 |
| | Computer | 2 |
| | TOTAL | 20 |
| Science | Physics | 3 |
| | Physics Experiment † | 2 |
| | Chemistry | 3 |
| | Chemistry Experiment † | 2 |
| | Biology | 3 |
| | Biology Experiment † | 2 |
| | Geology | 3 |
| | Geology Experiment † | 2 |
| | TOTAL | 20 |

† Includes computer use.

TABLE 5 Major Subject Studies Required for Mathematics and Science for Upper Secondary Teachers (as of April 2, 1991)

| Subject for Certificates | Major Subject Studies | Minimum Number of Credits |
|---|---|---|
| Mathematics | Algebra | 6 or 4 |
| | Geometry | 6 or 4 |
| | Analysis | 6 or 4 |
| | Probability & Statistics | 4 or 2 |
| | Computer | 4 or 2 |
| | TOTAL | 20 |
| Science | Physics | 4 |
| | Chemistry | 4 |
| | Biology | 4 |
| | Geology | 4 |
| | PH/CH/BI/GE Experiment† | 4 |
| | TOTAL | 20 |
| Agriculture | Agriculture-related Subj | 16 |
| | Career Guidance | 4 |
| | TOTAL | 20 |
| Technology | Technology-related Subj | 16 |
| | Career Guidance | 4 |
| | TOTAL | 20 |
| Career Guidance | Career Guidance | 4 |
| | Tech of Career Guidance | 10 |
| | Mgmt of Career Guidance | 6 |
| | TOTAL | 20 |

† Includes computer use.

SOURCE: Educational Personnel Certification Law (revised April 2, 1991).

Colleges and Universities," "Standards for the Establishment of Graduate Schools," "Standards for the Establishment of Junior Colleges," and "Standards for the Establishment of Technical Colleges." There is no certification system for teachers of higher institutions.

## SCHOOL CURRICULA FOR THE PROSPECTIVE SCIENCE AND TECHNOLOGY CAREER STUDENTS

### Current and Upcoming School Intended Curricula

The Japanese course of study, or national teaching guidelines, is determined by the MOE. The curriculum intended by the national government is revised about every 10 years (see Table 6) in a systematic, 5-step process:

1. Request to Curriculum Council (CC) by Education Minister

2. CC reports to Minister

3. MOE revises Course of Study

4. Course of Study implemented into schools

5. Begin appraisals to prepare for next revision

Total weekly class hours for grades 1-12 for recent years are shown in Table 7. Contrary to popular belief, Japanese yearly class hours are no longer than most of the other countries of Europe and North America, as shown in Table 7. Although yearly school days are 220 to 240, it does not mean that class hours are longer. Now that Japan has started to implement a plan for a 5-day school week, Japanese school days will be reduced to around 180, if the plan is completed.

Tables 8-11 show current and upcoming curricula of elementary, lower, and upper secondary schools of mathematics and science.

### School Textbooks: Mathematics and Science

Japanese textbooks are compiled based on the course of study determined and approved by the Ministry of Education. The use of authorized textbooks is mandatory at all school levels. All textbooks are pre-

**TABLE 6** Curricula Revision and Implementation in Recent Years

|  | Rev. → Imp. | Rev. → Imp. | Rev. → Imp. | Rev. → Imp. |
|---|---|---|---|---|
| Elementary | 1958  1961 | 1968  1971 | 1977  1980 | 1989  1992 |
| L. Secondary | 1958  1962 | 1969  1972 | 1977  1981 | 1989  1993 |
| U. Sec. G10 | 1960  1963 | 1970  1973 | 1978  1982 | 1989  1994 |
| U. Sec. G11 | 1960  1964 | 1970  1974 | 1978  1983 | 1989  1995 |
| U. Sec. G12 | 1960  1965 | 1970  1975 | 1978  1984 | 1989  1996 |

**TABLE 7** Weekly and Yearly Hours: G1-12 (1968, 1977, and 1989 revisions)

|  | Elementary |  |  |  |  |  | Lower Sec. |  |  | Upper Sec. |  |  |
|---|---|---|---|---|---|---|---|---|---|---|---|---|
| Grades | 1 | 2 | 3 | 4 | 5 | 6 | 7 | 8 | 9 | 10 | 11 | 12 |
| 1968 w. | 24, | 25, | 27, | 29, | 31, | 31; | 34, | 34*, | 34; | 34, | 34, | 34 |
| Rev. yr. | 816, | 875, | 945 | 1015, | 1085 | → | ← | 1190 | → | ← | 1190 | → |
| 1977 w. | 25, | 26, | 28, | 29, | 29, | 29; | 30, | 30, | 30; | 32, | 32, | 32 |
| Rev. yr. | 850, | 910, | 980 | ← | 1015 | → | ← | 1050 | → | ← | 1120 | → |
| 1989 w. | 25, | 26, | 28, | 29, | 29, | 29; | 30, | 30, | 30; | 32, | 32, | 32 |
| Rev. yr. | 850, | 910, | 980 | ← | 1015 | → | ← | 1050 | → | ← | 1120 | → |

NOTES:
(1) The number of teaching periods was reduced approximately 10 percent for the higher elementary grades and lower secondary schools, and 6 percent for upper secondary schools from 1977 revisions in order to avoid excessive study schedules.
(2) One school hour is 45 minutes for elementary schools and 50 minutes for secondary schools.
(3) * minimum schedule can be 33 or 1155 schools hours.
(4) Grade 1 is 34 weeks a year, but all others are 35 weeks a year.
(5) Upper secondary schedule is for full-time course only. Total teaching periods for part-time and correspondence courses (both three or more years, mostly four years) are the same as full-time courses, but they are spread into mostly four years.

pared by commercial publishers. Minor textbook revisions are carried out every three years, while major revisions are carried out when the course of study has been revised.

## CURRICULA ACTUALLY IMPLEMENTED BY THE TEACHER

Almost all elementary and lower secondary schools organize curriculum based on the course of study, or the national unified curriculum; therefore, little variation exists in terms of teaching content. Although class size and other factors differ from one school to the next, the Japanese government tries to give equal educational opportunities to all children, regardless of location, sex, race, creed, social status, economic position, or family origin. By shuffling teaching staff on a large scale and giving government subsidiaries to schools, the quality of teachers, as well as school facilities, does not vary much as far as public schools are concerned. There is one, and only one, public school within an attendance area. In addition, Japanese schools do not offer gifted/talented programs for each

**TABLE 8** Standard Number of Weekly and Yearly School Hours in Elementary Schools: Arithmetic and Science, 1980-1991 & Revised

|  | Grade |  | 1 | 2 | 3 | 4 | 5 | 6 |
|---|---|---|---|---|---|---|---|---|
| Arithmetic | 1980-91 Curr. | w. | 4 | 5 | 5 | 5 | 5 | 5 |
|  |  | yr. | 136 | 175 | 175 | 175 | 175 | 175 |
|  | 1992 Curr. | w. | 4 | 5 | 5 | 5 | 5 | 5 |
|  | (Revised) | yr. | 136 | 175 | 175 | 175 | 175 | 175 |
| Science | 1980-91 Curr. | w. | 2 | 2 | 3 | 3 | 3 | 3 |
|  |  | yr. | 68 | 70 | 105 | 105 | 105 | 105 |
|  | 1992 Curr. | w. | - | - | 3 | 3 | 3 | 3 |
|  | (Revised) | yr. | - | - | 105 | 105 | 105 | 105 |

NOTE: One school hour is 45 minutes.

**TABLE 9** Standard Number of Weekly and Yearly Hours in Lower Secondary Schools: Mathematics and Science, 1980-1992 vs. Revised (from 1993)

|  | Grade |  | 7 | 8 | 9 |
|---|---|---|---|---|---|
| Mathematics | 1989-92 Curr. | w. | 3 | 4 | 4 |
|  |  | yr. | 105 | 140 | 140 |
|  | 1993- Curr. | w. | 3 | 4 | 4 |
|  |  | yr. | 105 | 140 | 140 |
| Science | 1981-92 Curr. | w. | 3 | 3 | 3 |
|  |  | yr. | 105 | 105 | 140 |
|  | 1993- Curr. | w. | 3 | 3 | 3 or 4 |
|  |  | yr. | 105 | 105 | 105 or 140 |

NOTE: One school hour is 50 minutes.

**TABLE 10** Standard Number of Credits for General Education Subjects in Upper Secondary Schools: Mathematics and Science, 1982-1993

| Subject Areas | Course | Credit |
|---|---|---|
| Mathematics | Mathematics I | 4† |
|  | Mathematics II | 3 |
|  | Algebra and Geometry | 3 |
|  | Basic Analysis | 3 |
|  | Differential and Integral Calculus | 3 |
|  | Probability and Statistics | 3 |
| Science | Science I | 4† |
|  | Science II | 2 |
|  | Physics | 4 |
|  | Chemistry | 4 |
|  | Biology | 4 |
|  | Earth Science | 4 |

† Required.

**TABLE 11** Standard Number of Credits for General Education Subject in Upper Secondary Schools: Mathematics and Science, Revised

| Subject Areas | Course | Credit |
|---|---|---|
| Mathematics | Mathematics I | 4† |
|  | Mathematics II | 3 |
|  | Mathematics III | 3 |
|  | Mathematics A | 3 |
|  | Mathematics B | 2 |
|  | Mathematics C | 2 |
|  | Others | 2 |
| Science | Comprehensive Science | 4 |
|  | Physics I A | 2 |
|  | Physics I B * | 4 |
|  | Physics II | 2 |
|  | Chemistry I A | 2 |
|  | Chemistry I B * | 4 |
|  | Chemistry II | 2 |
|  | Biology I A | 2 |
|  | Biology I B * | 4 |
|  | Biology II | 2 |
|  | Earth Science I A | 2 |
|  | Earth Science I B * | 4 |
|  | Earth Science II | 2 |
|  | Others Related to Science |  |
|  | * Two subjects, four credits are required of all students |  |

† Required.

grade from first through twelfth.

The situation for upper secondary schools is very different from compulsory schools. As public schools are located in a large attendance area where several public schools are located, lower secondary graduates continued to upper secondary schools on the basis of selection from either entrance examinations or school recommendations. Thus, within one attendance area, several upper secondary schools are ranked according to students' academic abilities. Naturally, a wide variety of curricula is organized to meet the needs and abilities of the students.

## CURRICULA ATTAINED BY THE STUDENT

Deviation of average achievement scores of compulsory schools is strikingly small as far as public compulsory schools are concerned. The difference of average scores between urban and rural areas is also very small. On the other hand, large differences in average scores exist among upper secondary schools.

Achievement test scores of elementary children are characterized as an "inverted J curve." This means that the majority of children attain what is intended by the national government. In the case of lower secondary schools, the curve becomes somewhat flat. This implies that a large number of students do not attain the intended curriculum. The test scores of "Mathematics I," a requirement for all upper secondary students, for example, distributes along a letter U-shaped curve, which implies mathematic attainment of students is concentrated at two extremes. Some students do very well, while others either do not attain what is intended by the MOE or attain mediocre scores. In the Second International Mathematics Study, Japanese population B students, twelfth graders, of mathematics, science, or engineering fields attained high scores on an international standard, perhaps because the majority of these students belonged in the upper extreme of the curve.

Since upper secondary education of Japan has been universalized, and since student needs and abilities are so diverse, Japanese education is now heading toward individualization/personalization. The teaching-learning practices of the Japanese education scene are now drastically changing.

## STATISTICAL TRENDS OF SCIENCE AND TECHNOLOGY CAREER COURSES IN RECENT YEARS (1960-1990)

Science and technology (S&T) education in higher institutions is currently carried out through undergraduate courses in colleges and universities, junior colleges, graduate school, technical colleges, and special training schools. The latter two schools have been established rather recently.

Technical colleges were established in 1962 and designed to develop prospective human resources in S&T careers through a five-year training program for graduates of compulsory nine-year schooling. Departments and curricula of technical colleges have continually responded to the needs of the changing industrial structure. In 1992, the number of schools was 62 (54 national, 5 local/public, and 3 private), with spaces for 10,950 freshmen entrants (9,400 national, 920 local/public, and 603 private). The majority of graduates entered the industrial world; however, some graduates have continued their education at higher institutions. Fourteen percent of March 1991 technical college graduates went on to higher institutions.

Japan's main industry used to be agriculture. During the rapid economic expansion of the 1960s and 1970s, the demand for extensive human resource development (HRD) in the S&T fields was intensified by the business world. Special training schools were established in 1976 to supplement college education that offered vocational and technical education to meet the variety of needs of a changing society. Advanced courses in special training schools require graduation from an upper secondary institution, with courses designed for S&T careers. In May 1992, the total enrollment in advanced courses was 690,000, and 13 percent of March 1992 upper secondary graduates entered advanced courses. This number finally surpassed the number of entrants in junior colleges, which was 15.3 percent (520,000) of upper secondary graduates. Therefore, special training schools are now regarded as one of the important educational institutions for HRD in S&T fields.

The mainstream program for S&T careers is in the colleges, universities, and graduate schools. More and more prospective S&T students work on master's courses as the need for master's degree graduates becomes the preference of leading industrial firms.

To help present an overview of the HRD program for S&T careers, enrollment in science and engineering courses, or S&T-related courses, and of all higher education institutions in recent years is shown in Tables 12-26. These tables are labeled as Tables 12 through 14, Tables 15 through 17, Tables 18 through 20, Tables 21 through 23, Table 24, and Tables 25 and 26, and are related to four-year colleges and universities, junior colleges, graduate school master's courses, and graduate school doctoral courses, respectively. Tables 12, 15, 17, and 21 show the enrollment share in S&T fields by four-year colleges, junior colleges, graduate schools of master's courses, and graduate schools of doctoral courses, respectively. Tables 13, 16, 19, and 22 show the percentage of females enrolled in the S&T field by four-year colleges, junior colleges, graduate schools of master's courses, and graduate schools of doctoral courses, respectively. Tables 14, 17, 20, and 23 show the enrollment percentage by type of administration (national, public, and private) in the S&T fields by four-year colleges, junior colleges, graduate schools of master's courses, and graduate schools of doctoral courses, respectively. Table 24 shows statistics for technical colleges, and Tables 25 and 26 show statistics for special training schools.

# TRAINING SYSTEM IN THE BUSINESS WORLD

## HRD in Private Industry

Japanese private enterprises have developed a systematic training system in the last few decades. The system consists of three major fields: on-the-job-training, off-the-job-training, and self-development. Japanese higher education has not responded adequately to the needs of private industry. Japanese students work fairly hard until they enter colleges and universities; however, once students enter colleges and universities, they become complacent, and colleges and universities are not places of learning. A 1990 poll conducted by the Ministry of Education indicated that the average number of hours college students devote to their studies at home, in libraries, etc., other than during class periods, was just 6.3 hours a week for male students of humanities and social sciences, and 8.3 hours a week for female students. Although students in S&T fields are thought to work harder, the number of hours devoted to studying is just 9.2 hours for male students and 9.3 hours for female students. [Source: Waga Kuni no Bunkyou Shisaku Heisei 2-Nendo (Japanese Government Policies in Education, Science and Culture), 1990, pages 104-115, MOE.] As a result, the majority of Japanese enterprises do not expect much from higher education and try to establish their own educational system through on-the-job-training or other means.

Japanese education is universalized in terms of the qualitative expansion in the past 20 years. Currently, 38 percent of the graduates of upper secondary schools go on to colleges and universities, and, as of 1992, if we include the advanced courses of special training schools, 54 percent go on to higher education institutions. Japanese higher education has lots of room for improvement; however, the majority of Japanese people have a deep foundation of knowledge and skills of at least compulsory-level educational course content. Therefore, Japanese enterprises try to educate newly employed workers by their own efforts so that employees will develop to the expected level.

## Historical Trends of HRD in Private Industry

Japan implemented the job training system used by the United States, such as the Management Training Program and Training Within Industry, immediately after World War II. These programs include democratic ways of treating workers and establishing adequate working conditions. During the 1960s, Japan entered into the age of technological innovation and started to improve upon the American training system. These efforts established a lifetime employment convention quite different from the employment practices prior to World War II. The lifetime employment system enabled Japanese industry to establish an educational personnel division to train workers within their own schools, because they believed that trained workers will remain with their companies. Many colleges, or graduate-level schools, established by the enterprises have excellent educational programs comparable to top colleges and universities. Enterprises employ pioneering scientists and engineers as teachers.

Since Japan has entered the age of globalization, the Japanese management system is now being scrutinized by Western societies. However, with recent mobility of workers increasing, the job training system needs to be improved drastically in order to cope with the rapidly changing social and economic circumstances.

## Characteristics of the Japanese Job Training System

Since Japan implemented the American job training system, it is apparent that job training has been strongly influenced by the American mass production model. The Japanese convention is to employ workers for life, and it has become conspicuous to foster workers with a spirit to work for their own company. In addition, based on tradition and culture, it has become a common practice for elderly workers to teach younger workers while they are engaged in the daily routine of work. People learn willingly from elderly workers and do not show resistance to the customs. Since the majority of workers are literate, due to the success of basic education, it is not difficult to train workers in this manner.

Another characteristic to be noted is that self-development programs (SDP) have also become a common practice. Some companies even give money to those engaged in SDP on a reimbursement basis, or allow paid leave of absence or early leave for SDP activities.

Perhaps HRD programs in the business world have become one of the most important driving forces of Japan's rapid social, technical, and economic progress after World War II.

**TABLE 12** Percent of Enrollment in Institutions of Higher Education in Selected Major Fields of Study, 1960 to 1991

### UNDERGRADUATE COURSES

| BC | Year Japan Yr | | Humanities | Social Sciences | Science | Engineering | Agriculture | All Fields |
|---|---|---|---|---|---|---|---|---|
| 1960* | Showa | 35 | 12.9% | 42.9% | 2.7% | 15.4% | 4.7% | 601,464 |
| 1965* | | 40 | 12.7 | 43.1 | 3.0 | 19.5 | 4.1 | 895,645 |
| 1970 | | 45 | 12.7 | 41.8 | 3.1 | 21.1 | 3.7 | 1,344,358 |
| 1975 | | 50 | 13.1 | 41.7 | 3.0 | 20.2 | 3.5 | 1,652,003 |
| 1980 | | 55 | 13.8 | 40.5 | 3.1 | 19.4 | 3.4 | 1,741,504 |
| 1985 | | 60 | 14.2 | 38.7 | 3.4 | 19.8 | 3.5 | 1,734,392 |
| 1986 | | 61 | 14.4 | 38.6 | 3.4 | 19.9 | 3.5 | 1,758,635 |
| 1987 | | 62 | 14.5 | 38.8 | 3.4 | 19.8 | 3.4 | 1,806,024 |
| 1988 | | 63 | 14.7 | 39.1 | 3.3 | 19.8 | 3.4 | 1,861,306 |
| 1989 | Heisei | 1 | 15.1 | 39.4 | 3.3 | 19.7 | 3.4 | 1,929,137 |
| 1990 | | 2 | 15.2 | 39.6 | 3.4 | 19.6 | 3.4 | 1,988,572 |
| 1991 | | 3 | 15.4 | 39.8 | 3.4 | 19.6 | 3.3 | 2,052,335 |

NOTE: * Dates are based on a later edition.

**TABLE 13** Female Enrollment Ratio in Institutions of Higher Education in Selected Major Fields of Study, 1960 to 1991

### UNDERGRADUATE: PERCENTAGE OF FEMALE

| BC | Year Japan Yr | | Humanities | Social Sciences | Science | Engineering | Agriculture |
|---|---|---|---|---|---|---|---|
| 1960 | Showa | 35 | n/a | n/a | 11.8% | 0.5% | 1.5% |
| 1965 | | 40 | n/a | n/a | 12.4 | 0.4 | 3.0 |
| 1970 | | 45 | 52.3% | 5.2% | 13.4 | 0.6 | 5.9 |
| 1975 | | 50 | 59.7 | 7.8 | 14.4 | 0.9 | 8.8 |
| 1980 | | 55 | 58.3 | 8.2 | 15.9 | 1.5 | 11.5 |
| 1985 | | 60 | 59.4 | 9.3 | 18.1 | 2.7 | 14.7 |
| 1986 | | 61 | 60.6 | 9.9 | 18.1 | 2.9 | 15.0 |
| 1987 | | 62 | 62.0 | 10.6 | 17.9 | 2.9 | 15.5 |
| 1988 | | 63 | 63.9 | 11.7 | 18.1 | 3.1 | 17.0 |
| 1989 | Heisei | 1 | 64.9 | 13.0 | 18.3 | 3.4 | 18.8 |
| 1990 | | 2 | 65.9 | 14.6 | 18.5 | 3.9 | 20.9 |
| 1991 | | 3 | 66.4 | 16.0 | 19.0 | 4.7 | 23.9 |

SOURCE: Ministry of Education, Science and Culture. Statistical Abstract of Education, Science, and Culture, 1960 to 1992 edition.

**TABLE 14** Percent of Enrollment in National, Public, and Private Undergraduate Courses by Selected Major Fields of Study, 1966 to 1991

| Year | | | All Fields | | | Humanities | | | Social Sciences | | | Science | | | Engineering | | | Agriculture | | |
|---|---|---|---|---|---|---|---|---|---|---|---|---|---|---|---|---|---|---|---|---|
| BC | Jpn. | Yr | Natl | Pub | Priv | Natl | Pub | Priv | Natl | Pub | Priv | Natl | Pub | Priv | Natl | Pub | Priv | Natl | Pub | Priv |
| 1966 | Showa | 41 | 22.8 | 2.8 | 73.4 | 11.0 | 3.7 | 83.7 | 7.2 | 2.6 | 89.8 | 43.8 | 3.6 | 52.2 | 29.8 | 1.7 | 67.9 | 52.2 | 2.4 | 42.7 |
| 1970 | | 45 | 20.5 | 2.8 | 76.1 | 9.7 | 4.2 | 84.4 | 6.9 | 2.5 | 90.3 | 42.9 | 2.9 | 53.9 | 26.2 | 1.6 | 72.0 | 49.1 | 2.0 | 48.1 |
| 1975 | | 50 | 18.9 | 2.7 | 78.3 | 8.2 | 4.3 | 87.1 | 6.3 | 2.4 | 91.4 | 40.9 | 3.0 | 56.2 | 25.1 | 1.5 | 73.4 | 46.2 | 1.9 | 51.9 |
| 1980 | | 55 | 20.3 | 2.6 | 77.0 | 9.1 | 4.3 | 86.8 | 7.2 | 2.2 | 90.5 | 40.3 | 2.9 | 56.8 | 27.9 | 1.5 | 70.5 | 47.5 | 1.8 | 51.9 |
| 1985 | | 60 | 22.1 | 3.1 | 75.1 | 9.6 | 5.3 | 86.1 | 8.8 | 2.5 | 88.7 | 39.6 | 3.2 | 57.5 | 30.0 | 1.7 | 68.4 | 48.2 | 2.5 | 49.8 |
| 1986 | | 61 | 22.1 | 3.2 | 75.0 | 9.7 | 5.7 | 86.1 | 8.9 | 2.6 | 88.6 | 40.1 | 3.2 | 57.0 | 30.4 | 1.7 | 68.0 | 48.4 | 2.6 | 49.5 |
| 1987 | | 62 | 22.3 | 3.4 | 75.0 | 9.8 | 5.9 | 86.0 | 9.0 | 2.8 | 88.5 | 40.8 | 3.2 | 56.2 | 30.9 | 1.8 | 67.5 | 49.2 | 2.8 | 48.8 |
| 1988 | | 63 | 22.2 | 3.5 | 75.0 | 9.6 | 5.9 | 86.3 | 9.1 | 2.9 | 88.4 | 41.8 | 3.4 | 55.2 | 31.2 | 1.9 | 67.1 | 49.9 | 2.8 | 48.1 |
| 1989 | Heisei | 1 | 22.0 | 3.6 | 75.2 | 9.4 | 6.0 | 86.6 | 9.1 | 3.0 | 88.5 | 42.1 | 3.5 | 54.9 | 31.6 | 2.0 | 66.7 | 49.8 | 3.3 | 48.1 |
| 1990 | | 2 | 21.8 | 3.7 | 75.4 | 9.3 | 5.2 | 86.9 | 9.0 | 3.1 | 88.5 | 41.5 | 3.8 | 55.2 | 32.0 | 2.1 | 66.3 | 49.1 | 4.1 | 48.6 |
| 1991 | | 3 | 21.4 | 3.8 | 75.8 | 9.1 | 5.4 | 87.2 | 8.8 | 3.0 | 88.6 | 40.7 | 4.0 | 55.8 | 32.0 | 2.3 | 66.4 | 48.1 | 5.1 | 49.5 |

SOURCE: Ministry of Education, Science and Culture. Statistical Abstract of Education, Science, and Culture, 1967 to 1992 Edition.

KAZUO ISHIZAKA

**TABLE 15** Percent of Enrollment in Institution of Higher Education in Selected Major Fields of Study, 1960 to 1991

### UNDERGRADUATE COURSE

| BC | Year Japan Yr | | Humanities | Social Sciences | Science | Engineering | Agriculture | All Fields |
|---|---|---|---|---|---|---|---|---|
| 1960* | Showa | 35 | 17.9% | 19.8% | n/a | 11.3 | 1.7% | 81,528 |
| 1965* | | 40 | 21.8 | 15.2 | 0.1% | 10.2 | 1.3 | 145,458 |
| 1970 | | 45 | 19.8 | 11.6 | 0.1 | 8.4 | 1.3 | 259,757 |
| 1975 | | 50 | 21.1 | 10.9 | 0.0 | 6.7 | 1.2 | 348,922 |
| 1980 | | 55 | 21.3 | 9.1 | 1.9 | 5.5 | 1.1 | 366,248 |
| 1985 | | 60 | 23.2 | 9.8 | 2.3 | 5.4 | 1.1 | 366,180 |
| 1986 | | 61 | 24.3 | 10.3 | 2.3 | 5.3 | 1.1 | 391,078 |
| 1987 | | 62 | 25.0 | 10.9 | 2.4 | 5.2 | 1.0 | 432,393 |
| 1988 | | 63 | 25.3 | 11.6 | 2.5 | 5.3 | 0.9 | 444,808 |
| 1989 | Heisei | 1 | 25.4 | 12.2 | 2.7 | 5.1 | 0.8 | 455,696 |
| 1990 | | 2 | 25.8 | 12.7 | 3.0 | 5.0 | 0.8 | 473,194 |
| 1991 | | 3 | 26.2 | 13.2 | 3.3 | 5.0 | 0.8 | 497,559 |

Note: * Dates are based on a later edition.

**TABLE 16** Female Enrollment Ratio in Institutions of Higher Education in Selected Major Fields of Study, 1960 to 1991

### JUNIOR COLLEGES: PERCENTAGE OF FEMALE

| BC | Year Japan Yr | | Humanities | Social Sciences | Science | Engineering | Agriculture |
|---|---|---|---|---|---|---|---|
| 1960 | Showa | 35 | 81.2% | 12.4% | 100.0% | 1.7% | 14.7% |
| 1965 | | 40 | 91.2 | 21.3 | 78.5 | 3.3 | 18.3 |
| 1970 | | 45 | 96.6 | 44.0 | 88.7 | 3.3 | 12.8 |
| 1975 | | 50 | 97.8 | 54.7 | 97.0 | 3.8 | 15.1 |
| 1980 | | 55 | 99.1 | 63.5 | 97.6 | 8.8 | 18.0 |
| 1985 | | 60 | 98.0 | 69.7 | 98.0 | 15.3 | 22.9 |
| 1986 | | 61 | 98.1 | 73.0 | 98.5 | 17.3 | 25.6 |
| 1987 | | 62 | 98.4 | 76.2 | 98.8 | 19.2 | 26.5 |
| 1988 | | 63 | 98.4 | 77.4 | 98.9 | 20.2 | 28.0 |
| 1989 | Heisei | 1 | 98.4 | 78.3 | 99.3 | 22.9 | 30.9 |
| 1990 | | 2 | 98.4 | 79.5 | 99.6 | 26.2 | 34.8 |
| 1991 | | 3 | 98.2 | 80.1 | 99.6 | 29.3 | 37.8 |

SOURCE: Ministry of Education, Science and Culture. Statistical Abstract of Education, Science, and Culture, 1960 to 1992 edition.

**TABLE 17** Percent of Enrollment in National, Public, and Private Junior Colleges in Selected Major Fields of Study, 1966 to 1991

| Year | | | All Fields | | | Humanities | | | Social Sciences | | | Science | | | Engineering | | | Agriculture | | |
|---|---|---|---|---|---|---|---|---|---|---|---|---|---|---|---|---|---|---|---|---|
| BC | Jpn. | Yr | Natl | Pub | Priv | Natl | Pub | Priv | Natl | Pub | Priv | Natl | Pub | Priv | Natl | Pub | Priv | Natl | Pub | Priv |
| 1966 | Showa | 41 | 4.1 | 7.6 | 88.3 | 0.7 | 5.2 | 94.1 | 14.4 | 15.8 | 69.8 | 0.0 | 11.9 | 88.1 | 23.9 | 11.5 | 64.6 | 0.0 | 33.6 | 66.4 |
| 1970 | | 45 | 3.7 | 6.1 | 90.2 | 0.5 | 4.0 | 95.6 | 12.1 | 13.7 | 74.2 | 0.0 | 39.7 | 60.3 | 24.5 | 8.6 | 66.9 | 0.0 | 26.3 | 73.7 |
| 1975 | | 50 | 3.7 | 5.0 | 91.3 | 0.5 | 2.6 | 96.9 | 11.1 | 11.9 | 77.0 | 0.0 | 46.1 | 53.9 | 26.2 | 8.0 | 65.8 | 0.0 | 33.8 | 66.2 |
| 1980 | | 55 | 3.9 | 5.1 | 91.0 | 0.3 | 2.4 | 98.6 | 10.4 | 13.5 | 76.0 | 0.0 | 0.0 | 100.0 | 27.1 | 8.9 | 64.0 | 0.0 | 34.7 | 65.3 |
| 1985 | | 60 | 4.7 | 5.6 | 89.7 | 0.2 | 2.6 | 97.3 | 11.1 | 12.4 | 76.5 | 0.0 | 0.0 | 100.0 | 26.3 | 7.3 | 66.4 | 0.0 | 35.4 | 64.6 |
| 1986 | | 61 | 4.6 | 5.2 | 90.2 | 0.1 | 2.3 | 97.5 | 10.2 | 11.3 | 78.5 | 0.0 | 0.0 | 100.0 | 26.1 | 5.6 | 68.4 | 0.0 | 36.6 | 63.4 |
| 1987 | | 62 | 4.3 | 4.8 | 90.9 | 0.1 | 2.0 | 97.9 | 8.9 | 9.9 | 81.2 | 0.0 | 1.3 | 98.7 | 24.7 | 3.9 | 71.4 | 0.0 | 37.8 | 62.2 |
| 1988 | | 63 | 4.2 | 4.9 | 90.9 | 0.1 | 1.9 | 97.9 | 8.1 | 9.4 | 82.5 | 0.0 | 2.7 | 97.3 | 23.1 | 3.9 | 73.0 | 0.0 | 39.8 | 60.2 |
| 1989 | Heisei | 1 | 4.1 | 4.8 | 91.1 | 0.1 | 2.0 | 97.9 | 7.6 | 9.1 | 83.3 | 0.0 | 2.8 | 97.2 | 21.0 | 3.9 | 75.2 | 0.0 | 37.8 | 62.2 |
| 1990 | | 2 | 3.8 | 4.7 | 91.5 | 0.1 | 1.9 | 98.9 | 7.0 | 8.5 | 84.5 | 0.0 | 2.8 | 97.2 | 17.4 | 3.9 | 78.7 | 0.0 | 33.4 | 66.6 |
| 1991 | | 3 | 3.5 | 4.4 | 92.0 | 0.1 | 1.8 | 98.0 | 6.4 | 7.7 | 85.9 | 0.0 | 2.7 | 97.3 | 13.5 | 3.6 | 82.8 | 0.0 | 32.8 | 67.2 |

SOURCE: Ministry of Education, Science and Culture. Statistical Abstract of Education, Science, and Culture, 1960 to 1992 Edition.

CAREERS IN SCIENCE AND TECHNOLOGY: AN INTERNATIONAL PERSPECTIVE

**TABLE 18** Percent of Enrollment in Institutions of Higher Education in Selected Major Fields of Study, 1960 to 1991

### GRADUATE SCHOOL MASTER'S COURSES

| BC | Year Japan Yr | | Humanities | Social Sciences | Science | Engineering | Agriculture | All Fields |
|---|---|---|---|---|---|---|---|---|
| 1960 | Showa | 35 | 34.6% | 28.5% | 11.9% | 14.7% | 4.5% | 8,305 |
| 1965 | | 40 | 18.5 | 20.0 | 13.1 | 33.7 | 6.1 | 16,771 |
| 1970 | | 45 | 18.6 | 16.6 | 10.8 | 37.0 | 7.4 | 27,714 |
| 1975 | | 50 | 17.8 | 13.7 | 9.6 | 40.3 | 8.0 | 33,560 |
| 1980 | | 55 | 15.3 | 11.3 | 10.5 | 41.5 | 7.1 | 35,781 |
| 1985 | | 60 | 11.7 | 9.1 | 9.5 | 42.9 | 10.2 | 48,147 |
| 1986 | | 61 | 11.3 | 9.1 | 9.8 | 43.5 | 9.8 | 51,094 |
| 1987 | | 62 | 10.8 | 9.2 | 9.9 | 43.9 | 10.1 | 54,352 |
| 1988 | | 63 | 10.5 | 9.5 | 10.3 | 45.1 | 8.4 | 56,596 |
| 1989 | Heisei | 1 | 10.2 | 9.9 | 10.6 | 46.0 | 6.6 | 58,228 |
| 1990 | | 2 | 9.7 | 10.3 | 10.5 | 45.9 | 6.5 | 61,884 |
| 1991 | | 3 | 9.2 | 10.6 | 10.2 | 46.1 | 6.6 | 68,739 |

SOURCE: Ministry of Education, Science and Culture. Statistical Abstract of Education, Science, and Culture, 1960 to 1992 edition.

**TABLE 19** Female Enrollment Ratio in Institutions of Higher Education in Selected Major Fields of Study, 1976 to 1991

### GRADUATE SCHOOL MASTER'S COURSES: PERCENTAGE OF FEMALE

| BC | Year Japan Yr | | Humanities | Social Sciences | Science | Engineering | Agriculture |
|---|---|---|---|---|---|---|---|
| 1976 | Showa | 51 | 27.0 | 8.6 | 7.5% | 0.5% | 5.1% |
| 1980 | | 55 | 31.9 | 12.5 | 8.5 | 1.1 | 7.7 |
| 1985 | | 60 | 33.3 | 18.4 | 8.9 | 1.8 | 14.7 |
| 1986 | | 61 | 35.6 | 18.4 | 8.9 | 2.1 | 16.5 |
| 1987 | | 62 | 37.3 | 20.2 | 9.6 | 2.4 | 17.8 |
| 1988 | | 63 | 39.1 | 21.4 | 10.2 | 2.6 | 15.5 |
| 1989 | Heisei | 1 | 41.8 | 23.3 | 11.0 | 3.0 | 10.8 |
| 1990 | | 2 | 43.9 | 24.7 | 12.1 | 3.3 | 11.3 |
| 1991 | | 3 | 46.3 | 27.9 | 12.7 | 3.7 | 12.9 |

SOURCE: Ministry of Education, Science and Culture. Statistical Abstract of Education, Science, and Culture, 1960 to 1992 edition.

**TABLE 20:** Percent of Enrollment in National, Public, and Private Master's Courses in Selected Major Fields of Study, 1976 to 1991

| | Year | | All Fields | | | Humanities | | | Social Sciences | | | Science | | | Engineering | | | Agriculture | | |
|---|---|---|---|---|---|---|---|---|---|---|---|---|---|---|---|---|---|---|---|---|
| BC | Jpn. | Yr | Natl | Pub | Priv | Natl | Pub | Priv | Natl | Pub | Priv | Natl | Pub | Priv | Natl | Pub | Priv | Natl | Pub | Priv |
| 1976 | Showa | 51 | 57.0 | 4.0 | 39.0 | 26.3 | 3.4 | 70.2 | 17.0 | 2.9 | 80.1 | 75.1 | 4.8 | 20.1 | 70.2 | 3.4 | 26.4 | 86.2 | 5.2 | 86 |
| 1980 | | 55 | 61.7 | 3.6 | 34.7 | 27.7 | 3.6 | 68.7 | 22.6 | 2.7 | 74.8 | 77.1 | 4.1 | 18.8 | 75.0 | 2.9 | 22.1 | 86.3 | 4.4 | 92 |
| 1985 | | 60 | 62.8 | 3.7 | 33.5 | 30.0 | 3.6 | 66.4 | 26.7 | 3.6 | 69.7 | 73.2 | 4.2 | 22.6 | 73.3 | 2.8 | 23.9 | 64.0 | 4.3 | 318 |
| 1986 | | 61 | 62.8 | 3.6 | 33.6 | 31.4 | 3.6 | 65.0 | 26.7 | 3.6 | 69.7 | 72.1 | 4.0 | 23.8 | 72.6 | 2.8 | 24.5 | 64.9 | 3.9 | 311 |
| 1987 | | 62 | 63.0 | 3.7 | 33.3 | 31.4 | 3.8 | 64.8 | 27.9 | 3.4 | 68.7 | 71.8 | 3.9 | 24.3 | 72.5 | 2.9 | 24.6 | 66.4 | 3.4 | 302 |
| 1988 | | 63 | 63.3 | 3.7 | 33.0 | 31.2 | 3.7 | 65.1 | 27.8 | 3.4 | 68.8 | 70.3 | 4.0 | 25.7 | 71.6 | 2.9 | 25.4 | 74.2 | 3.5 | 223 |
| 1989 | Heisei | 1 | 63.8 | 3.8 | 32.4 | 32.1 | 3.6 | 64.3 | 28.2 | 3.7 | 68.2 | 70.1 | 4.5 | 25.4 | 71.0 | 3.0 | 14.8 | 86.0 | 3.5 | 105 |
| 1990 | | 2 | 63.8 | 3.9 | 32.3 | 33.1 | 3.3 | 63.6 | 29.0 | 4.4 | 66.6 | 71.5 | 4.6 | 23.9 | 70.4 | 3.1 | 26.5 | 86.1 | 3.3 | 107 |
| 1991 | | 3 | 63.5 | 3.8 | 32.7 | 32.7 | 3.3 | 64.0 | 29.9 | 4.1 | 66.0 | 71.6 | 4.7 | 23.8 | 69.5 | 3.1 | 27.4 | 85.8 | 3.6 | 106 |

SOURCE: Ministry of Education, Science and Culture. Statistical Abstract of Education, Science, and Culture, 1960 to 1992 Edition

**TABLE 21** Percent of Enrollment in Institutions of Higher Education in Selected Major Fields of Study, 1960 to 1991

### GRADUATE SCHOOL DOCTOR'S COURSE

| BC | Year Japan Yr | | Humanities | Social Sciences | Science | Engineering | Agriculture | All Fields |
|---|---|---|---|---|---|---|---|---|
| 1960* | Showa | 35 | 13.7% | 12.0% | 12.1% | 5.3% | 4.6% | 7,429 |
| 1965* | | 40 | 11.0 | 9.3 | 10.7 | 11.0 | 3.6 | 11,683 |
| 1970 | | 45 | 14.2 | 13.0 | 17.1 | 17.8 | 6.3 | 13,243 |
| 1975 | | 50 | 16.5 | 14.7 | 15.8 | 16.9 | 6.8 | 14,904 |
| 1980 | | 55 | 15.7 | 13.3 | 14.2 | 12.9 | 6.0 | 18,211 |
| 1985 | | 60 | 15.0 | 11.3 | 11.5 | 11.2 | 5.1 | 21,541 |
| 1986 | | 65 | 14.3 | 10.7 | 10.9 | 12.2 | 5.3 | 23,177 |
| 1987 | | 61 | 13.4 | 10.3 | 10.9 | 13.0 | 5.4 | 24,562 |
| 1988 | | 62 | 13.0 | 9.8 | 10.9 | 14.1 | 5.7 | 25,880 |
| 1989 | Heisei | 1 | 12.8 | 9.6 | 11.0 | 14.3 | 5.7 | 27,035 |
| 1990 | | 2 | 12.7 | 9.4 | 10.8 | 15.2 | 6.1 | 28,354 |
| 1991 | | 3 | 12.2 | 9.2 | 11.0 | 16.3 | 6.5 | 29,911 |

Note: *Dates are based on a later edition.

SOURCE: Ministry of Education, Science and Culture. Statistical Abstract of Education, Science, and Culture, 1960 to 1992 edition.

**TABLE 22** Female Enrollment Ratio in Institutions of Higher Education in Selected Major Fields of Study, 1976 to 1991

### GRADUATE SCHOOL DOCTOR'S COURSES: PERCENTAGE OF FEMALE

| BC | Year Japan Yr | | Humanities | Social Sciences | Science | Engineering | Agriculture |
|---|---|---|---|---|---|---|---|
| 1976 | Showa | 51 | 17.3% | 6.0% | 5.0% | 1.3% | 4.9% |
| 1980 | | 55 | 23.3 | 7.0 | 5.2 | 2.1 | 6.8 |
| 1985 | | 60 | 27.4 | 12.3 | 6.8 | 3.0 | 10.1 |
| 1986 | | 61 | 27.5 | 13.8 | 7.5 | 3.8 | 11.1 |
| 1987 | | 62 | 28.0 | 14.5 | 7.3 | 4.5 | 11.7 |
| 1988 | | 63 | 29.1 | 15.7 | 7.6 | 5.1 | 13.2 |
| 1989 | Heisei | 1 | 30.5 | 16.9 | 8.2 | 5.5 | 14.7 |
| 1990 | | 2 | 31.7 | 18.8 | 8.4 | 5.3 | 13.8 |
| 1991 | | 3 | 34.1 | 20.9 | 9.5 | 5.5 | 13.1 |

SOURCE: Ministry of Education, Science and Culture. Statistical Abstract of Education, Science, and Culture, 1960 to 1992 edition.

**TABLE 23** Percentage of Enrollment in National, Public, and Private Doctor's Courses in Selected Major Fields of Study, 1976 to 1991

| Year | | | All Fields | | | Humanities | | | Social Sciences | | | Science | | | Engineering | | | Agriculture | | |
|---|---|---|---|---|---|---|---|---|---|---|---|---|---|---|---|---|---|---|---|---|
| BC | Jpn. | Yr | Natl | Pub | Priv | Natl | Pub | Priv | Natl | Pub | Priv | Natl | Pub | Priv | Natl | Pub | Priv | Natl | Pub | Priv |
| 1976 | Showa | 51 | 59.8 | 6.3 | 33.9 | 35.8 | 4.2 | 60.0 | 38.9 | 7.7 | 53.3 | 83.4 | 5.7 | 10.8 | 73.6 | 4.6 | 21.9 | 86.2 | 5.7 | 8.1 |
| 1980 | | 55 | 58.5 | 6.0 | 35.6 | 38.9 | 4.6 | 56.5 | 38.1 | 6.1 | 55.8 | 81.2 | 6.0 | 12.7 | 75.4 | 3.7 | 20.8 | 88.7 | 3.4 | 7.9 |
| 1985 | | 60 | 59.4 | 5.8 | 34.8 | 41.2 | 4.4 | 54.4 | 40.5 | 5.7 | 53.8 | 82.2 | 6.5 | 11.3 | 80.5 | 2.5 | 17.0 | 89.1 | 4.1 | 6.8 |
| 1986 | | 61 | 60.7 | 5.5 | 33.8 | 41.3 | 4.0 | 54.8 | 41.2 | 5.9 | 52.9 | 82.9 | 6.1 | 11.0 | 81.3 | 2.5 | 16.2 | 89.6 | 3.4 | 7.0 |
| 1987 | | 62 | 62.0 | 5.4 | 32.6 | 41.2 | 4.1 | 54.8 | 41.7 | 6.2 | 52.2 | 82.8 | 5.6 | 11.6 | 83.0 | 2.0 | 15.0 | 89.7 | 3.4 | 6.9 |
| 1988 | | 63 | 63.0 | 5.4 | 31.6 | 42.2 | 4.1 | 53.8 | 41.2 | 6.4 | 52.5 | 82.3 | 5.5 | 12.2 | 84.1 | 2.2 | 13.7 | 90.4 | 3.1 | 6.5 |
| 1989 | Heisei | 1 | 63.9 | 5.3 | 30.8 | 42.2 | 4.9 | 53.0 | 41.8 | 6.2 | 52.0 | 83.0 | 5.1 | 11.9 | 84.6 | 1.9 | 13.4 | 91.7 | 2.6 | 5.7 |
| 1990 | | 2 | 64.9 | 5.3 | 29.8 | 43.2 | 5.2 | 51.6 | 42.6 | 6.0 | 51.5 | 83.2 | 4.9 | 11.9 | 85.4 | 1.8 | 12.8 | 91.9 | 2.4 | 5.7 |
| 1991 | | 3 | 65.5 | 5.2 | 29.2 | 42.7 | 4.9 | 52.4 | 42.1 | 6.3 | 51.6 | 84.7 | 4.6 | 10.7 | 85.1 | 1.9 | 13.0 | 90.9 | 2.6 | 6.5 |

SOURCE: Ministry of Education, Science and Culture. Statistical Abstract of Education, Science, and Culture, 1960 to 1992 Edition.

**TABLE 24** Enrollment in National, Public, and Private Technical Colleges, 1962 to 1991

| BC | Year Japan | Yr | Total | Female | National | Local | Private | Percentage of Female |
|---|---|---|---|---|---|---|---|---|
| 1962 | Showa | 37 | 3,375 | 35 | 1,549 | 703 | 1,123 | 1.0 |
|  | 1962 |  | (1.0%) | (45.9%) | (20.8%) | (33.3%) |  |  |
| 1965 |  | 40 | 22,208 | 347 | 14,839 | 2,920 | 4,449 | 1.6 |
| 1966 |  | 41 | 28,795 | 484 | 20,206 | 3,529 | 5,060 | 1.7 |
|  | 1966 |  | (1.7%) | (70.2%) | (12.3%) | (17.6%) |  |  |
| 1970 |  | 45 | 44,314 | 673 | 33,091 | 3,919 | 7,304 | 1.5 |
| 1971 |  | 46 | 46,707 | 649 | 34,928 | 3,881 | 7,898 | 1.4 |
|  | 1971 |  | (1.4%) | (74.8%) | (8.3%) | (16.9%) |  |  |
| 1975 |  | 50 | 47,955 | 736 | 38,194 | 3,942 | 5,819 | 1.5 |
| 1976 |  | 51 | 47,055 | 761 | 38,417 | 3,972 | 4,666 | 1.6 |
|  | 1976 |  | (1.6%) | (81.6%) | (8.4%) | (9.9%) |  |  |
| 1980 |  | 55 | 46,348 | 917 | 39,211 | 4,018 | 3,119 | 2.0 |
| 1981 |  | 56 | 46,468 | 1,017 | 39,320 | 4,030 | 3,118 | 2.2 |
|  | 1981 |  | (2.2%) | (84.6%) | (8.7%) | (6.7%) |  |  |
| 1985 |  | 60 | 48,288 | 1,723 | 40,739 | 4,148 | 3,401 | 3.6 |
| 1986 |  | 65 | 49,174 | 2,023 | 41,597 | 4,140 | 3,437 | 4.1 |
|  | 1986 |  | (4.1%) | (84.0%) | (8.4%) | (7.0%) |  |  |
| 1987 |  | 61 | 50,078 | 2,432 | 42,543 | 4,160 | 3,375 | 4.9 |
| 1988 |  | 62 | 50,934 | 2,997 | 43,486 | 4,145 | 3,303 | 5.9 |
| 1989 | Heisei | 1 | 51,966 | 3,753 | 44,612 | 4,131 | 3,223 | 7.2 |
| 1990 |  | 2 | 52,930 | 4,677 | 45,627 | 4,126 | 3,177 | 8.8 |
| 1991 |  | 3 | 53,698 | 5,856 | 46,436 | 4,190 | 3,072 | 10.9 |
|  | 1991 |  | (10.9%) | (86.5%) | (7.8%) | (5.7%) |  |  |

SOURCE: Ministry of Education, Science and Culture. Statistical Abstract of Education, Science, and Culture, 1960 to 1992 edition.

**TABLE 25** Number of Graduates of Special Training Schools by Type of Courses, 1977 to 1991

| BC | Year Japan | Yr | Total | Female | Upper Sec Course | Advanced Course | General Course | Percentage of Female |
|---|---|---|---|---|---|---|---|---|
| 1977 | Showa | 52 | 95,997 |  | 16,471 | 66,381 | 13,145 |  |
| 1978 |  | 53 | 177,465 |  | 26,005 | 131,039 | 20,421 |  |
| 1980 |  | 56 | 208,669 |  | 31,486 | 159,716 | 17,467 |  |
| 1984 |  | 59 | 247,882 | 145,953 | 34,003 | 173,324 | 40,555 |  |
|  |  |  | (female | 55.9% | 72.3% | 62.6% | 31.3%) |  |
| 1985 |  | 60 | 262,716 |  | 33,590 | 183,553 | 45,573 |  |
| 1989 | Heisei | 1 | 333,025 |  | 41,123 | 230,835 | 61,067 |  |
| 1990 |  | 2 | 350,360 | 177,707 | 42,522 | 247,960 | 59,878 |  |
|  |  |  | (female | 51.9% | 62.8% | 52.3% | 27.9%) |  |
| 1991 |  | 3 | 366,603 | 185,421 | 42,110 | 266,222 | 58,271 |  |
|  |  |  | (female | 50.6% | 61.7% | 51.3% | 26.9%) |  |

SOURCE: Ministry of Education, Science and Culture. Statistical Abstract of Education, Science, and Culture, 1984, 1991, and 1992 Editions.

**TABLE 26** Enrollment of Special Training Schools by Type of Control, 1976 to 1991 with Number of Schools in Selected Years

| BC | Year Japan Yr | Total | Female | National | Local | Private | Percentage of Female |
|---|---|---|---|---|---|---|---|
| 1976 | Showa 51 | 131,492 | 104,425 | 3,481 | 4,641 | 123,370 | 79.4 |
| | (schools) | ( 893 | | 46 | 28 | 819 = 91.7%) | |
| 1977 | 52 | 356,790 | 256,918 | 15,952 | 11,774 | 392,064 | 72.0 |
| | (schools) | ( 1,941 | | 192 | 80 | 1,669 = 86.0%) | |
| 1978 | 53 | 416,438 | 285,881 | 15,714 | 18,615 | 382,109 | 68.6 |
| | (schools) | ( 2,253 | | 190 | 114 | 1,949 = 86.5%) | |
| 1980 | 55 | 432,914 | 287,938 | 15,843 | 20,628 | 396,443 | 66.5 |
| | (schools) | ( 2,520 | | 187 | 146 | 2,187 = 86.8%) | |
| 1985 | 60 | 538,175 | 312,185 | 18,070 | 24,069 | 496,036 | 58.0 |
| | (schools) | ( 3,015 | | 178 | 173 | 2,664 = 88.4%) | |
| 1986 | 65 | 587,609 | 332,312 | 18,127 | 25,549 | 543,933 | 56.6 |
| 1987 | 61 | 653,026 | 359,209 | 18,119 | 26,023 | 608,884 | 55.0 |
| 1988 | 62 | 699,534 | 357,762 | 18,013 | 26,112 | 655,409 | 53.7 |
| 1989 | Heisei 1 | 741,682 | 390,871 | 17,548 | 26,849 | 746,193 | 52.7 |
| 1990 | 2 | 791,431 | 410,543 | 17,433 | 27,805 | 746,193 | 51.9 |
| | (schools) | ( 3,300 | | 166 | 182 | 2,952 = 89.5%) | |
| 1991 | 3 | 834,710 | 424,817 | 17,453 | 28,599 | 788,661 | 50.9 |
| | (schools) | 3,370 | | 163 | 185 | 3,022 = 89.7%) | |
| | (graduates) | 366,603 | 185,421 | 6,455 | 10,650 | 349,498 | |

SOURCE: Ministry of Education, Science and Culture. Statistical Abstract of Education, Science, and Culture, 1984, 1991, and 1992 editions.

## REFERENCES

Ishizaka, K. School Education in Japan (1989). Tokyo: International Society for Educational Information.

Ministry of Education, Science, and Culture (MOE). Waga Kuni no Bunkyou Shisaku Shisaku Heisei 1-Nendo to 4-Nendo (Japanese Government Policies in Education, Science and Culture 1989 ed. to 1992 ed.) Tokyo: MOE.

Ministry of Education, Science, and Culture. Statistical Abstract of Education, Science, and Culture. 1992 ed. (Published annually) (1958 to 1992). Tokyo: MOE.

# Trends in Science and Technology Careers: Education Through Research

## Dervilla Donnelly

## INTRODUCTION

In Europe, scientific and technological research has developed within national, political, and economic frameworks. The need to strengthen European industrial competitiveness in world markets is now influencing the direction of scientific and technological research and, to an extent, the training mode of researchers.

Today, skills and knowledge constitute two of the few areas where an economy can amend the differential competitive advantage; thus, education and training are critical elements of policy affecting society. By policy I mean a statement of goals for the training and maintaining of a scientific and technological workforce, the means by which such manpower can be achieved, and the motivation and mobilization of such a workforce.

## NEW CHALLENGES FOR SCIENCE EDUCATION IN EUROPE

To attain a more highly-qualified workforce, democratic governments have promoted an expansion in higher education based on the principle of equality of opportunity, and have made science education more relevant to life in the community (Husén, Tuijnman, and Halls). The questions being asked are:

- What is the role of science in the school teaching curriculum?

- What is its effect on the development of student skills and societal skills?

- What is the value of nurturing an awareness of sciences?

- Will this cause a lowering of standards by dismissing academic skills?

The last question is not a totally acceptable fear, for the space created will open the doors to biotechnology, environmental studies, and information technology—the societal skills related to understanding the applications of science. The International Council of Associations for Science Education (ICASE) has initiated *Project 2000+. Toward Science and Technology Education for All—A Basic Human Need?* (Holbrook, 1992). This project plans to:

- clearly identify ways of promoting the development of scientific and technological literacy for all;

- put forward educational programs in such a way as to empower all to satisfy their basic needs and be productive in an increasingly technological society;

- provide guidelines for the continuous professional development of science and technology educators and leaders;

- encourage the formation of national task forces to initiate local programs for greater scientific and technological literacy;

- support the development of a wide range of projects that aim to improve quality of life and productivity in society; and

- support the evaluation of existing and projected programs to ensure scientific and technological literacy goals are being met.

## SKILL SHORTAGE IN EUROPE

Current and anticipated skill shortages are a threat to European competitiveness. The decreasing supply of young—though more highly-qualified—persons and the pace of technological change emphasize the need for a heavy investment in upgrading the existing workforce. To be considered urgently is a reassessment of educational policies and greater support for continuing education. Every person should have at least some basic understanding of science and technology, and no European should leave school unqualified and unskilled.

Some assistance in solving the skill shortages can be achieved by additional training of the existing science and technology workforce.

- For governments, this implies a shift in their priorities for education. There should be a raising of the basic level of education and arrangements for continuing education.

- For industry, it will mean retraining and updating employees to improve the competitiveness of companies.

- For universities and higher education institutions, a continuing educational program should become a mainstream activity as well as an acceptance of a more structured cooperation with industry.

- For individuals, it will mean the acceptance of the need to update on a regular basis. Interestingly, there are already 158 University Enterprise Training Partnerships (UETP): regional, sectoral, and mixed. The regional UETPs are training in all technology sectors of companies and institutions in their geographical area. Sectoral UETPs concentrate on a particular type of technology and its applications, while mixed UETPs are relevant to both aspects (see Figure 1) (IRDAC Opinion).

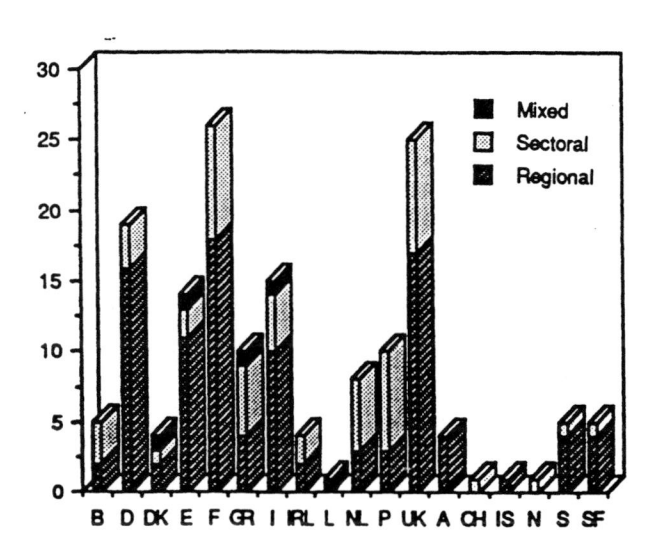

**FIGURE 1** UETPs in the EC and EFTA member states, 1990.

## REDRESSING THE BALANCE

In the short-term and medium-term, it is evident that the needs of industry cannot be met by the production of new graduates. Therefore, the need to enable qualified scientists and engineers who have left their profession for any reason to return at a level appropriate to their earlier training and experience became a requirement. Such returners could include graduates who have left engineering or science in order to work in accountancy or commerce, but who might be encouraged to return to a creative role in research and development or in marketing high-technology products. Loss of status must be avoided.

A major group of potential returners are women who have left their profession for family reasons. The number in this capacity will increase in the future. The European Commission has stressed the necessity to exploit this potential and has committed to supporting projects in scientific and technological research. This substantial commitment must be matched by an equally vigorous policy of investment in the communities' human resources.

It is fully accepted that existing and anticipated human resource requirements for new technology at an advanced level clearly exceeds current higher education output, both quantitatively and qualitatively. This situation is true in the United States and in Japan, and

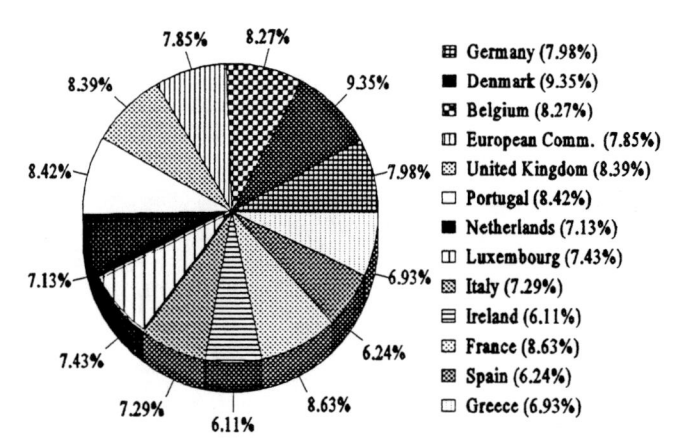

**FIGURE 2** Females as percentage of total EC labor force.

Germany (7.98%)
Denmark (9.35%)
Belgium (8.27%)
European Comm. (7.85%)
United Kingdom (8.39%)
Portugal (8.42%)
Netherlands (7.13%)
Luxembourg (7.43%)
Italy (7.29%)
Ireland (6.11%)
France (8.63%)
Spain (6.24%)
Greece (6.93%)

industry considers that the lack of qualified people represents a major obstacle to full exploitation of new technologies.

Of the working population in Europe, one worker in three is female, 38.4 percent. In the UK the figure is even higher, above 40 percent. Forty-three percent of married women are members of the working population (see Figure 2). In most countries there is only a small percentage of women in the engineering and technological areas (IRDAC Opinion).

Typical European Science Foundation projects include training for women in micro-electronics and technical skills in the building industry, and in accountancy training. Spain, Portugal, and Greece have been very successful in attracting funds from the European Science Foundation.

## WOMEN IN SCIENCE AND TECHNOLOGY

The underrepresentation of women in some areas of science education is an unacceptable waste of intellectual and economic resources. A gap still exists in 1993, but its closure is a major educational challenge. "If I were a king," wrote Madame du Chatelet in the eighteenth century, "I would redress an abuse which cuts back half of mankind. I would have women participate in all human rights, especially those of the mind." With a few notable exceptions, women have played a secondary role to men in the world of scientific discovery.

Technological subjects at the higher education level have the lowest proportion of women students compared to any other field of study. Figures suggest that a disproportionately low number of girls take science and technology options in school. There are more girls than boys in secondary education in the UK, Belgium, France, Ireland, Germany, and Luxembourg. Only after A Levels, or the equivalent, and above the age of 20, do women become the minority. At the age of 24, there are twice as many boys as girls still in education. There are great differences between men and women in some disciplines at further educational levels as well (see Figure 3).

It is apparent that in most European countries, boys and girls still go down very different paths when it comes to preparing for a future career. The demographic changes of an aging population, and a dramatic fall in young people entering the labor market, will mean that women will be of increasing importance as a force of labor. It is predicted that in the year 2000, 44 percent of the labor force will be women. It is a belief that it will be necessary to eliminate explicit and implicit discrimination against women and recognize and provide for the particular problems facing women returning to work.

The European Science Foundation and the Equal Opportunities Commission are encouraging women to go into traditional male occupations and new technologies.

The Women in Technology in the European Community (WITNEC)-UETP was established in 1988 to try to address educational and training issues. It is a network of partners working for the motivation, development, and support of women in science, technology, and enterprise. WITNEC partners include representatives from industry and enterprise, promoters who are all interested in the formulation, orientation,

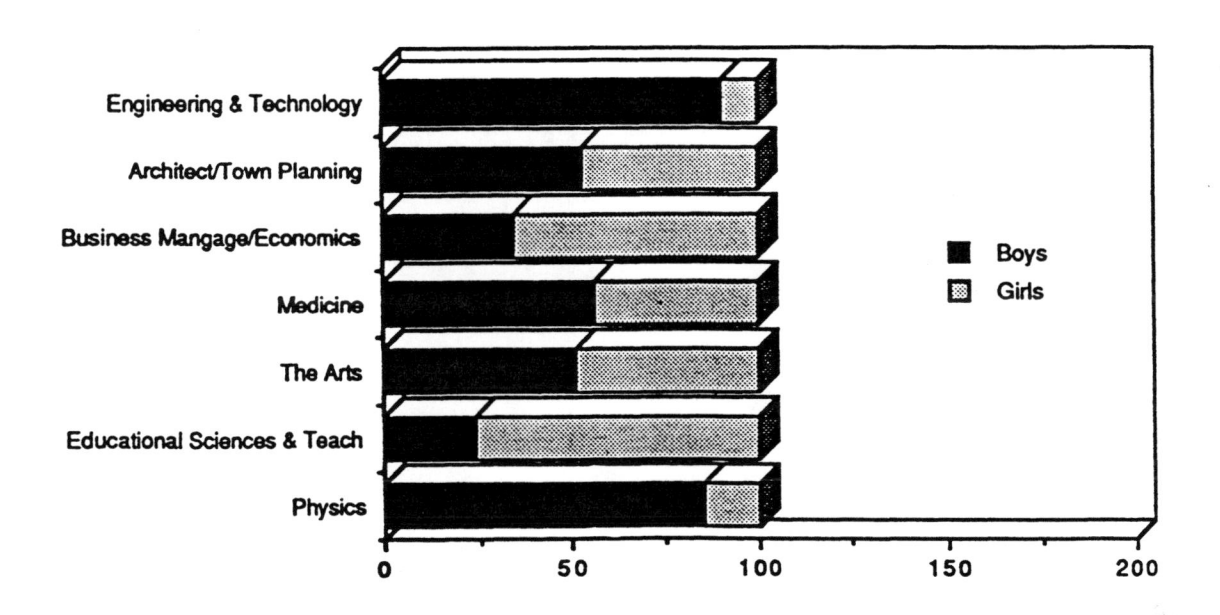

**FIGURE 3** Education in Europe.

and research on women in technology and are drawn from all the EC and EFTA member states. WITNEC seeks to publicize educational programs aimed at increasing and diversifying the range of subject choices available to girls and to facilitate the entry of young adults into career paths and jobs in the field of information technology, electronics, and other technology-related industries.

The WITNEC-UETP aims to:

- generate a network of universities, enterprises, and other organizations committed to increasing the number of girls and women taking up studies, careers, and autonomous activities in advanced technological fields across the EC and EFTA states;

- assist women studying and working in technological fields to surmount barriers that interfere with their entering into the labor market and support their technological career development;

- promote and support research concerning problems in this field and their solutions and to disseminate the results;

- publicize problems, strategies, and methods already found for dealing with them, especially within the overall COMETT network;

- generate action programs and pilot new initiatives of a general or specific character;

- encourage women undertaking technological careers to contribute to overcoming reluctance to employing women and at the same time promote information and experience exchanges between universities and enterprises;

- improve women's foreign language and interpersonal skills;

- take women with basic math and scientific knowledge, often in low-level jobs, through programs to the point where they could enter vocational education courses and ultimately higher level courses;

- offer knowledge and skill development in technological fields to women without any previous technological background via a program of evaluation of their skills and characteristics, and offer tools to facilitate this;

- contribute to the creation of a favorable learning environment for women; and

- test the content and procedure of existing resources on the theme "girls and women in science technology and enterprise," by collecting and exchanging information on research, statistics, policies, and action projects.

## ADVANCING EUROPEAN COOPERATION

### European Science Foundation

Without doubt there is ample evidence of demand for a stronger scientific and technological workforce. How can this be supplied, and what part will research play in this educational process? By networking the existing scientific and technological workforce, one progresses toward a stronger Europe. Umberto Columbo, President of the European Science Foundation (ESF), has stated: "In difficult times the advantages of resources sharing and international cooperation become all the more obvious. By these means each member organization can avoid unnecessary duplication, share common costs and equipment, pool ideas and talents in joint actions."

The ESF was founded to promote excellent basic science in Europe. Its aim was to advance European cooperation in basic research by planning, launching, and, where appropriate, managing collaborative research programs that help its member organizations (research councils and academies) to achieve their own aims and objectives. It aimed to promote mobility to build scientific communities on a European scale in specific fields and to enhance scientific competence in Europe.

The success of the ESF in bringing together the natural and social sciences has been impressive, particularly in those areas—such as the study of environmental matters—where scientific research is vital in providing adequate and accurate evidence to underpin economic and political decisions affecting the world's population. In line with the many challenging changes that have taken place in the wider Europe of the last few years, and continue to take place, the nature of ESF support for basic science has changed and, perhaps, become responsive to the demands of scientists in the Europe that is now taking shape.

There is a plethora of European and American programs of scientific meetings that have varied philosophies, aims, and organizational patterns. The aims and philosophies of some of these programs are often not sufficiently recognized. At its meeting in Oslo in June 1989, the Executive Council of the ESF decided to promote a continuing program of research conferences in Europe. A proposal has been submitted to the Executive Council from the presidents of the science organizations of the Federal Republic of Germany.

### Research Conferences

The concept of research conferences is best exemplified by the well-known U.S. Gordon research conferences that have evolved over 60 years into a powerful instrument in U.S. research activities in chemistry and some related disciplines. These conferences were initiated in the late 1920s by Dr. Neil E. Gordon of Johns Hopkins, who was aware of the many problems and difficulties in establishing good direct communications between scientists working in particular areas, and sometimes in different disciplines of science. He proposed that these research conferences be held in secluded locations, the groups small but of a highly-qualified character, the discussions informal and off-the-record, and the specification of the scientific subject at the frontier of knowledge. Presently, there are some 130 Gordon conferences per year attended by about 15,000 scientists from countries all around the world.

### European Research Conferences

A shared scientific understanding leading naturally to scientific communications, cooperation, and exchange has in the past provided an important bridge between European nations. This bridge has often been built many years ahead of other economic and political boundaries. The exchange of ideas and knowledge has contributed to the historical process of creating a more coherent European identity. More specifically, such scientific interchanges will speed up the pace of scientific discoveries and contribute to a more rational use of scarce resources, human effort, talent, and money.

The European research conferences have two aims of equal importance:

1. The scientific need to provide a framework for the scientific debate that is the essential component of truly innovative research, the result of the collision of ideas that often occurs in scientific argument.

2. The need to build a sense of Europe-wide identity, especially among young researchers—not Europe-against-the-world but European rather than the narrow nationalistic view.

These conferences will give the young researchers an opportunity to meet the established leaders in their fields, as well as colleagues of their own generation from all over Europe and the world. The ESF believes that these conferences will offer a stimulating environment for scientific argument and an opportunity for the young researchers. These conferences were not to be kept as purely for academe. It was hoped that they would attract researchers from industry, as well as researchers in particular from Central and Eastern European countries. An emphasis was, of course, also put on the hope that American scientists would be interested in these conferences, as over the years the Americans have generously invited European scientists to the U.S. Gordon research conferences.

The foundation has proposed that their program of European research conferences cover all scientific disciplines, and, in due course, expand the conferences into the social sciences and humanities. This direction of expansion is quite a challenge as these research conferences have evolved more in the natural sciences.

## Program of EUROSCO

The European research conference (EUROSCO) program consists of a series of scientific discussion meetings. Each series is devoted to the same general subject and normally takes place about every other year. The core activity at such research conferences is based on invited lectures by leading scientists in the field, followed by extensive discussion. The conferences are held in carefully-selected locations conducive to facilitating the interaction of participants over a five-day period. In order to encourage speakers to present their latest results and ideas, no written papers are requested and there are no conference proceedings. To facilitate participants associating freely and establishing new professional contacts that often lead to new collaborative research, the number of participants is restricted, with an upper limit of 100. In addition to five hours per day of formal lectures and discussion poster sessions, round-table discussions or groups are often arranged in the evening. A plenary session is normally held toward the end of the conference, in which an attempt is made to assess the scientific value of the meeting and also to discuss the scientific orientation of the next two conferences in the series or if the conference line is to be continued. This procedure results in the selection of topics for the conferences to be recommended, or within the context of the conferences themselves, and then sent to the ESF, which makes the formal decision whether to proceed or not. At the plenary session, the vice-chairperson for the next conference, and thus the chairperson for the conference after that, is also recommended. The conference chairman and the organizing committee are responsible only for the scientific input. The management of the conference is under the umbrella of the ESF office.

At present the European research conferences are not self-supporting and are funded by the EC. The Gordon conferences are self-supporting through a non-profit organization incorporated in New Hampshire as a voluntary corporation for scientific purposes.

There are other meetings besides the Gordon conferences that have very definite style, for example the NATO meetings.

## Other Science Programs

### The NATO Science Program

The NATO Science Program was established in 1958 in recognition of the crucial role of science and technology in maintaining the economic, political, and military strength of the Atlantic community. Together with the projects designed to enhance the quality of life undertaken by the committee on the challenges of modern society, it forms what is sometimes referred to as the "third dimension" of the North Atlantic alliance, the non-military dimension concerned with the enhancement of contact between member nations in the areas of science and technology, culture, and the problems of modern society.

All participants are expected to play an active role

at the meeting, and they are invited by their contributions to research and the potential response of the meeting.

### *Jacques Monod Conferences*

The Life Sciences Department of CNRS (Centre Nationale de la Recherche Scientifique) organizes a series of 10 conferences per year called the Jacques Monod conferences. The topics chosen are usually recent progress obtained in the different domains of fundamental biology and its applications in biotechnology, health, agronomy, and their associated industries. The total number of people attending these conferences is restricted to a maximum of 60-65. The objective is to offer those attending the opportunity and time for in-depth discussions with all their colleagues in order to facilitate organizing collaborations between teams and laboratories from different countries. Presentations must be recent unpublished work or work in the process of being published.

### *The CIBA Conferences*

The aim of the CIBA Conference Foundation in holding conferences is to foster international cooperation. The most important consideration is scientific quality. Eight or nine conferences are held each year, and twenty to thirty participants, all of whom must be active in the field, are invited. Topics may be proposed by the CIBA Foundation itself, or by others, and are then filtered through the international peer review system. A recent evaluation of the conferences by the CIBA Foundation showed that the publication of the conference papers appeared to have an impact in the wider scientific community in that they are cited early, frequently, and for a long period of time.

Also in Europe, there is a EUCHEM program in chemistry, which is now being run under the auspices of the ESF. EuroNet conferences in engineering science, which have a long tradition, are also run through the Foundation secretariat. EMBO research workshops have similar characteristics to the EUCHEM program, with the advantage of a more solid funding.

## NETWORKS

Coupled with Euroconferences are the many network schemes that are currently in vogue in Europe. The idea behind networks is simple but fundamental. All over Europe there are scientists and scholars active in the same field of research and addressing the same kind of scientific questions, all with their own projects and facilities. Networks aim to bring these researchers together by offering them a platform where they can discuss their activities and develop plans for future collaborations. The main ideas of the networks and conferences are to foster mutual awareness, to promote mobility in building scientific communities on a European scale, and to facilitate interdisciplinary research.

### ESF Networks

ESF Networks have the following characteristics:

- Scientific topics and activities are proposed "bottom up" by groups of active scientists. Major decisions about the scientific orientation of the networks are made by the coordination committee for that network, composed of the leading scientists involved. The scientists also take a leading role in the management and administration of each network, although support from the ESF office is sometimes available.

- The topics selected for networks (chosen on the basis of originality and excellence) are usually at a critical stage at which a significant step forward can be achieved through European collaboration. Interdisciplinarity is also an important feature in many networks.

- Member organizations of the Foundation and the ESF standing committees can influence the direction of the network scheme in two ways: first, by encouraging scientists to present proposals in particular topics, and second, by advising the chairmen of the standing committees about particular proposals.

- Networks operate for a period of three years with a total budget in the range of five hundred to eight hundred thousand FF. The core activity is usually a series of three or four workshops for twenty to thirty persons. Other activities may include smaller working meetings on individual subtopics of the network, exchanges for both senior and younger scientists, planning of joint research activities, publications, periodic newsletters to inform the relevant scientific community, and preparation of research inventories and databases.

- Every network is, on completion, evaluated by a group of independent experts appointed by the Network Committee.

### The Human Capital and Mobility Program

The Human Capital and Mobility Program covers all scientific and technological sectors such as mathematics and information science, physics, chemistry, life science, engineering, earth science, and the environment.

The program also covers areas of social and human sciences likely to improve European competitiveness and bring about sustainable economic development in fields such as economic and management sciences, environmental economics, and in the interconnections between science, technology, and society, and to deal with the general public's understanding and acceptance of science and technology.

There are four main activities carried out under the program:

1. *Fellowships*.

2. *Networks*. Scientific and technical cooperation networks are being created and developed across Europe, paying special attention to the needs of the less favored regions. This activity aspires to develop research networks linking several teams or laboratories whose capacity is complementary. These links will boost the effects of the European Community's research programs in specific areas (see Figure 4).

3. *Large-Scale Facilities*.

4. *Euroconferences*. The organization of a series of high-level meetings at the cutting edge of scientific and technical knowledge is intended to strengthen the cohesion of the EC by giving young researchers the chance to come into contact with and benefit from high levels of expertise in specific science and technology areas (see Figure 5).

The overall participation in the Human Capital and Mobility Program is impressive. Statistics on the 1992 program are given in Figure 6.

### REFERENCES

Colombo, U. 1993. European Science Foundation Communication, March 1993.

Holbrook, J.B. 1992. Project 2000+. Towards Science and Technology Education for All—A Basic Human Need? Science International Newsletter, September-December 1992, p.45.

Husén, T. (ed), A. Tuijnman, and W.D. Halls. Schooling in Modern European Society. A report of the Academia Europa. Pergamon Press.

IRDAC Opinion. Industrial Research and Development Advisory Committee of the Commission of the European Communities. Skills Shortages in Europe.

Women in Technology in the European Community University Enterprise Training Partnership, p. 10-11.

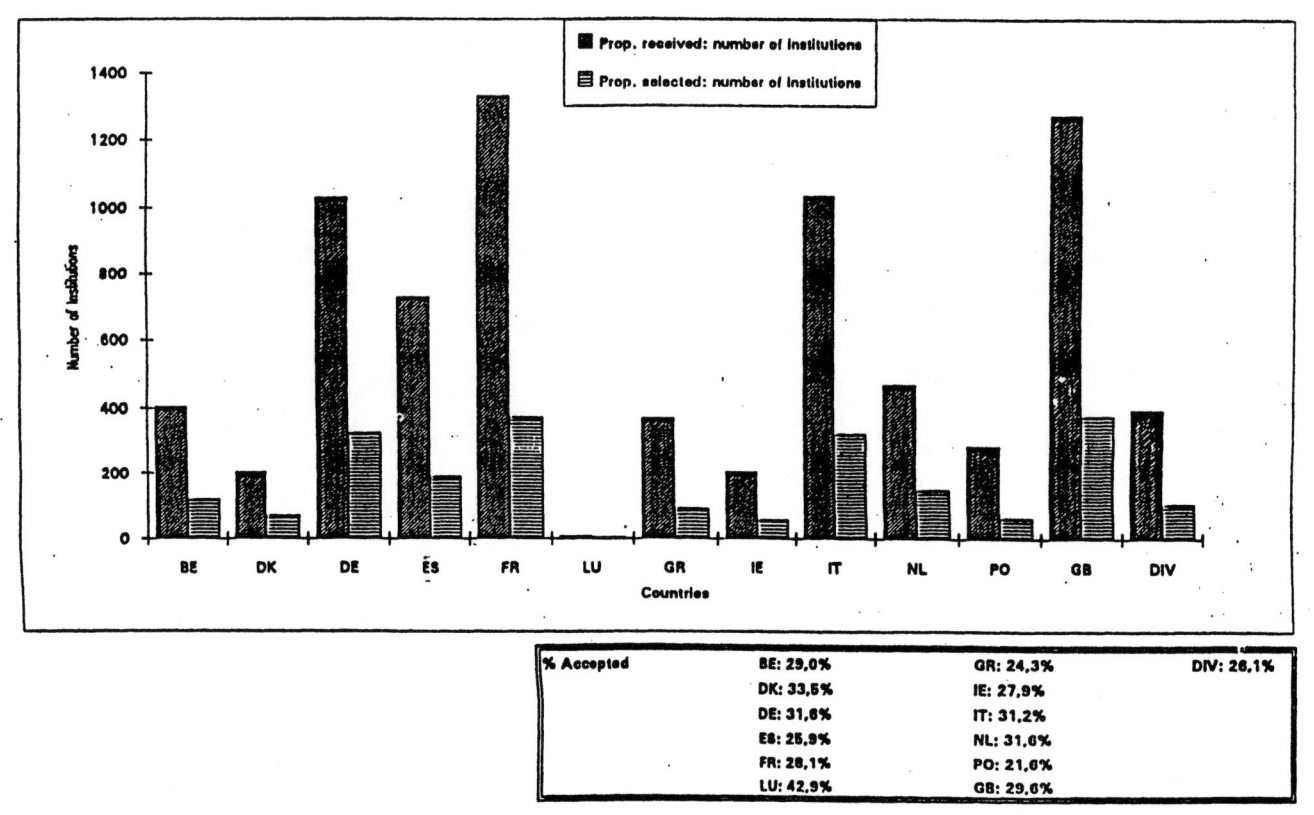

**FIGURE 4** Networks 1992: participating countries.

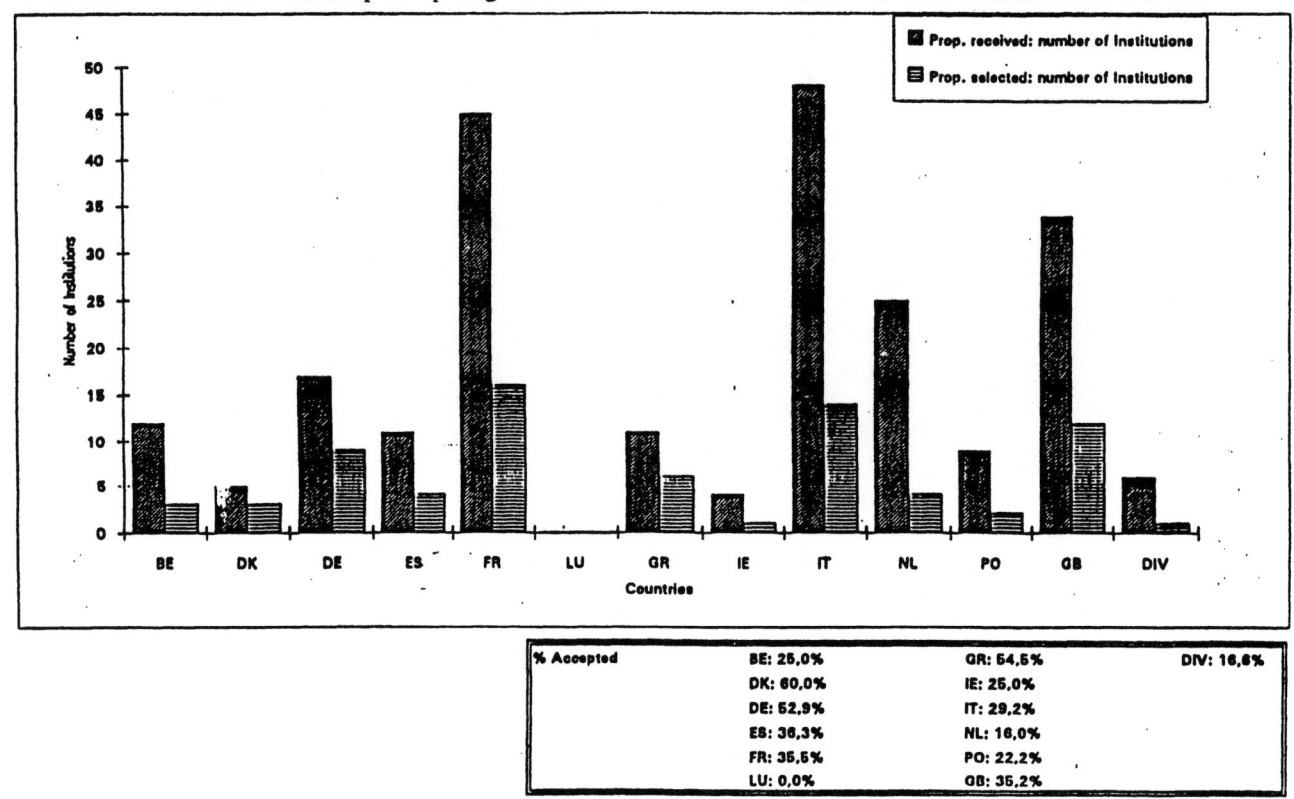

**FIGURE 5** Euroconferences 1992: participating countries.

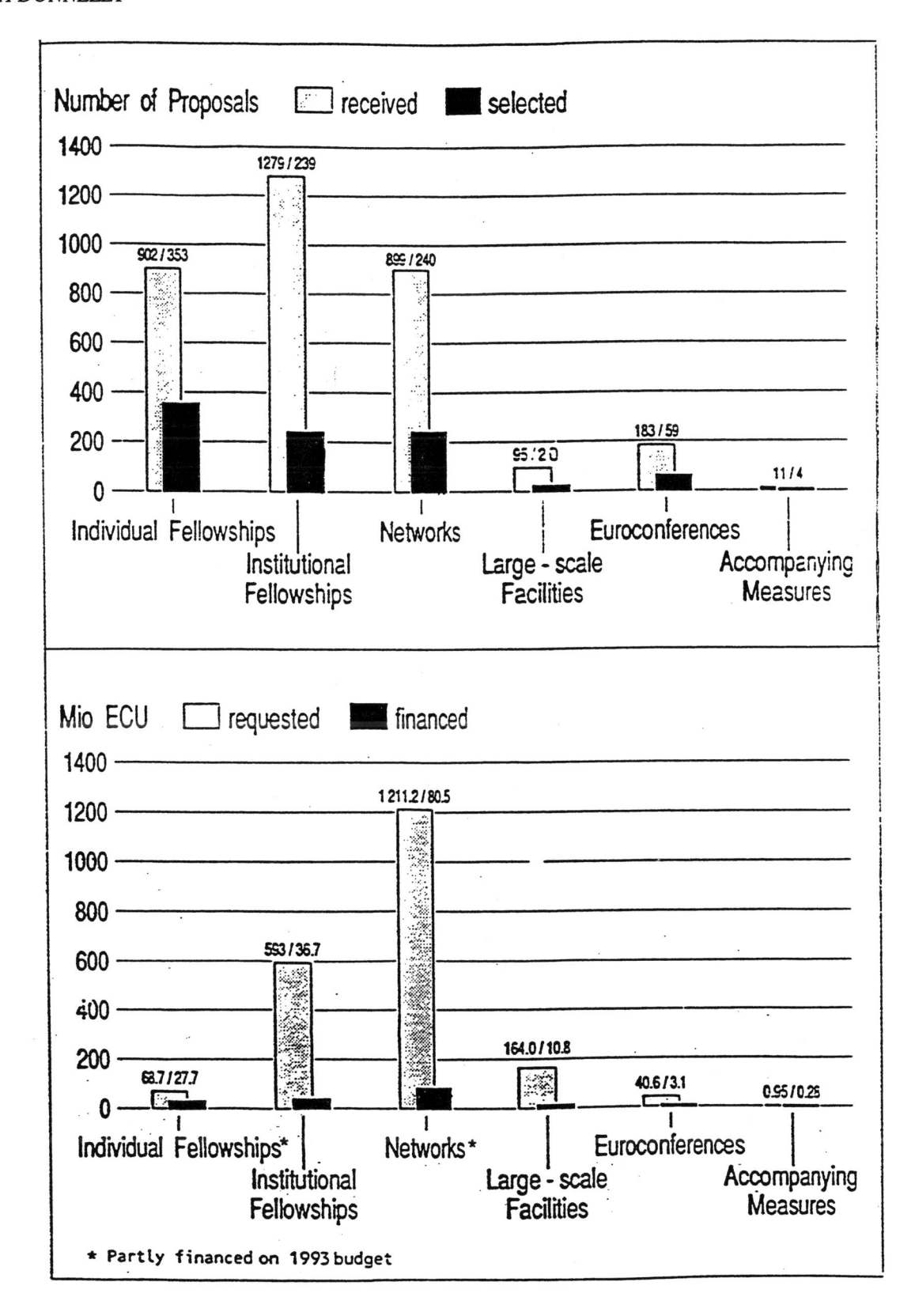

**FIGURE 6** Human Capital and Mobility 1992: overall participation.

# Utilizing Points of Intervention: A Critique

## Pamela Ebert Flattau

Over the past two days we have explored the concept of a career in science and technology (S&T), those factors thought to influence an individual's decision to enter a career, and key career stages. We identified a number of programs aimed at promoting S&T careers. However, it is clear from the studies presented at this conference that more fundamental information is needed about the development of the S&T career before fully effective human resource policies and programs can be designed.

We explored the possibility that statistical databases maintained in a number of countries would be useful for understanding S&T career patterns. After all, most countries with a serious commitment to S&T support the collection of statistics about the S&T workforce. We found, however, that available data sets are generally more useful for tracking changes in the size and composition of the labor force than for documenting the factors that influence the development of S&T careers. That is, these data sets tell us about the movement of individuals from job to job but little about the determinants of career change—such as problems of skill obsolescence or changes in career goals.

Changes will be needed, therefore, in national and international data collection practices to provide information about S&T careers. That is, data are needed that go beyond the question:

*Have we been successful in increasing the size of the S&T workforce?*

to answer the question:

*Have our programs and policies been successful in recruiting talented individuals into S&T and launching them into productive careers?*

Introducing changes in national data collection efforts to monitor S&T careers is, however, a serious undertaking that merits considerable thought and careful attention to issues of design. Consider, for example, that quantitative experts at this conference have described many factors involved in constructing national and international data sets that simply track the number of individuals entering or leaving the pool of workers in S&T! Furthermore, as John Moore reminded us yesterday, coupled with econometric models, labor force statistics in most countries have served human resource policies quite well over the years. Thus, there is a kind of inertia inherent in the extension of existing statistical systems to include career tracking due to the success of their past performance in monitoring the S&T labor force.

The subject of this conference springs, however, from a growing concern within the science community that a career in S&T has become less attractive to young people than in the past despite government efforts to foster interest in this area. Given the limitations of current data sets, how are we to persuade national leaders to invest the resources that are needed to monitor S&T careers? A first step is to identify the types of information that should be collected, and three areas suggest themselves:

1. Career exploration and career planning as it occurs in educational settings

2. The nature of the transition from school to work

3. Career advancement in the work setting

## CAREER EXPLORATION AND PLANNING IN EDUCATIONAL SETTINGS

Many nations have developed strategies to recruit individuals into S&T careers by focusing on improvements in the educational environment. Emphasis has been given to the development of new curricula, to teacher preparation, and to increased stipend support for talented individuals to pursue advanced studies in S&T. Dr. Ishizaka has described for us the significant changes Japan has introduced in its education system in recent years. These reforms address the preparation of teachers and curriculum for the universal preparation of Japanese students at the primary and secondary school levels.

The success of educational reforms in science and mathematics is often measured in terms of changes in overall student performance on standardized mathematics and science tests, especially in the international arena (e.g., relative ratings in the International Association for the Evaluation of Educational Achievement, or IEA). According to Dr. Ishizaka, Japan is also monitoring the success of their educational reforms based on changes in these performance measures.

But the ultimate goal of many of these reforms is to direct individuals—especially talented individuals—into S&T careers. To monitor progress toward that goal measures other than performance measures are needed. Retention studies are suggested that monitor the number of students at various points along the education path so that conclusions may be drawn about the efficacy of educational reforms relative to the fraction of individuals making the transition from one stage of the educational process to the next.

While studies that monitor the number of individuals in the education path are useful and important, they are often predicated on the belief that career development is a linear process and that changes in the educational environment will contribute to the flow of individuals through the system. In fact, as Dr. Miller told us earlier in this meeting, career planning is far more dynamic than many linear models predict and, furthermore, education is but one factor (albeit a critical factor) in the evolution of the S&T career.

Longitudinal studies are also needed to deepen our understanding of the S&T career process within the context of the education system. Such studies have the potential of documenting critical phases of career development and of suggesting intervention strategies within the education system that might assure the recruitment and retention of qualified individuals into S&T.

A number of speakers at this meeting, such as Paul Baltes, Yu Xie, Jon Miller, and Thomas Whiston, have identified some aspects of career exploration and career planning that might be examined through longitudinal analyses:

- Goal setting: levels of aspiration, timing of career decisions, students' understanding of occupations, students' understanding of career options

- Attitudes: toward S&T, toward mathematics and science studies, toward the life of scientists

- Achievement or performance: on standardized tests in science and mathematics, differences between students who aspire to S&T careers and those who aspire to other vocations

- Parental expectations and parental education

- Gender differences in the exploration of S&T career options

- Environmental variables

Such studies might be of special interest to analysts in Japan as they head toward further educational reforms that will emphasize "individualization/personalization" of S&T studies at the secondary school level.

## SCHOOL-TO-WORK TRANSITIONS

Another promising area for study is the stage of "school-to-work transitions." Dr. Donnelly has offered an especially useful example of an intervention activity that seems to have been designed with specific sensitivity to a critical stage of career development: the time when young investigators have completed their advanced training and may need assistance in establishing themselves among the larger research community. The European Research Conference, which Dr. Donnelly described, aims to: (1) provide a framework for scientific debate, and (2) build a sense of European-wide identity, especially among young researchers.

Carefully designed career outcomes studies would be especially useful in documenting the success of this program. A properly designed outcome study might focus on such questions as: What are the near-term and long-term effects of the Euroconference in launching young investigators into productive careers? Did the individual take the next step in the career process as a result of this program? Did this program enhance the level of productivity of the worker?

Coupled with longitudinal studies of career development, program outcome studies have the potential of making significant contributions to the ways in which educators, policymakers, and planners will approach the development of intervention strategies in the coming years—whether through curricular reform or through informal strategies like those described by Dr. Donnelly.

## CAREER STAGES IN THE WORK SETTING

Earlier in the meeting, Dr. Jaworski observed that many corporations have attempted to define the career path of employees and have introduced methods to sustain the professional competency of their staffs. Many companies will allow highly trained scientists, for example, to have a certain amount of free time for their own research while conducting research activities contributing to corporate goals. In some industries, "dual ladder" programs have been developed with a scientific advancement component and a managerial advancement component. While no industry uses all these—or other—techniques, almost all companies use some combination of these approaches.

The paucity of data on adult career development in S&T in work organizations is striking. Few attempts have been made to evaluate the effects of programs such as those described by Dr. Jaworski either on the employee's career goals or on those of the organization. Specialized studies that emphasize mid- to late-career development in S&T are sorely needed. Such information would be especially useful for identifying exemplary programs that sustain research and development skills and programs that successfully facilitate transition through critical stages of work-to-retirement transitions.

## CONCLUSION

At the outset of the conference, Richard Pearson reminded us that there are many aspects of the concept of a career that must be addressed before statistical systems can be designed to track the development of S&T careers. This includes the need for a common definition of an S&T career, some understanding of the key determinants of a career, and key transition stages that make tracking possible.

Fortunately, sufficient groundwork has been laid by the social and behavioral science communities to begin planning for the systematic collection of information about the development of the S&T career, although much work remains before we have sorted through competing theoretical perspectives to determine which offer more promise for informing human resource policy and planning activities. This conference has provided us with an important opportunity to consider what types of information might be collected in the coming years.

# PART V

# Influencing Science and Technology Career Trends: The Role of International Organizations

# Profiles of Participating Organizations

## ACADEMIA EUROPAEA

The Academia Europaea is an international association of individual scholars having as its prime aim the promotion of education, learning, and research. The scope includes the humanities; law; economic, social, and cognitive sciences; mathematics; medicine; and all branches of the natural and technological sciences. Its objectives are to promote and support excellence in European scholarship, research, and education; further the development of a European identity in scholarship and research; act as a European center for scholars; and conduct interdisciplinary and international studies and research.

The Academia has initiated a number of study groups. The first of these has published a major study of school education in Europe and reported to the Council of Europe's Education Committee. A second study is about to report on psychosocial problems of youth. Ongoing studies include higher education and the teaching of science in schools.

In addition to these studies on training, education, and development, an extremely useful three-day meeting was held in 1991 on the evolution of academic institutions and research councils. The discussions have been summarized in a report published by the Academia, where the problems of Central and Eastern European academies were highlighted. Aid to Central and Eastern European countries has been given by the establishment of, for example, 20 prizes for young Russian scientists in the humanities, biology, and mathematical sciences.

As a contribution to the intellectual formation of European scientists, Academia Europaea has begun publication of the quarterly *European Review,* which will cover a wide range of topics including human capital development.

The Academia's activities are clearly filling an important gap in European scholarship, with particular emphasis on the cross-fertilization that results from both an interdisciplinary and an international approach, in areas often not treated or sponsored elsewhere. The value of the Academia is beginning to be widely appreciated both in academic and in other circles throughout Europe. The topics indicated here show a range of interest as strong in the humanities and the social sciences as in the natural sciences. Two forthcoming meetings are planned that will set the cultural background and policy considerations on the "Idea of Progress" and the "Integrated Approach to Science and Technology Policy." The latter will be concerned mainly with the incorporation of social science considerations into science policymaking.

## COMMISSION OF THE EUROPEAN COMMUNITIES

The European Commission is the civil service of the European Communities. Its powers extend to preparing and proposing and then to executing and monitoring European Community legislation. The legislation is, however, approved and accepted by the member states through the Council of Ministers, sometimes on the basis of co-decision and after consulting the European Parliament.

In spite of the emphasis on subsidiary in the Maastricht Treaty, the number of different fields in which the Communities can intervene has greatly increased from the starting point of the coal and steel and the nuclear sectors to include agriculture and fisheries, industry, employment, transport, energy, social affairs, research, training, education, etc.

Therefore, within the Commission, there are many different groups actively working on the subject of human resources in science and technology and related issues such as trends in science and technology careers. Many parts of the Commission have specific initiatives in this area:

- The Human Capital and Mobility Program operated by the Directorate General for Science, Research, and Development directly addresses these issues as part of the Community's main funding program for research and technological development, the Framework Program.

- Training elements exist within each of the specific sectorial research programs that make up the Framework Program. These programs are operated by the Directorates General dealing with science, research and development, industry, telecommunications, agriculture, fisheries, and transport.

- There are the various student exchange programs (e.g., COMETT and ERASMUS) operated by the Task Force for Human Resources, Education, Training, and Youth.

- The part of the Commission dealing with employment, industrial relations, and social affairs is also very interested and active in this area.

- As part of the worldwide drive to improve statistics and indicators on human resources in science and technology, the Statistical Office of the European Communities, with the Directorate General for Science, Research, and Development, OECD, the United States, and Canada, is leading the work on the development of a methodological manual for statistics and indicators in the field of human resources in science and technology and is also making pilot collections of data on a regional basis.

The Commission is not only tackling the problem of human resources in science and technology by addressing the training needs that result directly from new developments in science and technology, but it is also investigating new tools that would enhance human competence, support existing employment, and improve the quality of life.

## EUROPEAN SCIENCE FOUNDATION

The European Science Foundation (ESF) was established in 1974. Its member organizations are the major European academies, research councils, and other institutions supporting scientific research nationally. Although these organizations are funded mainly by governments, the ESF itself is a nongovernmental organization. It maintains close associations with other international bodies with interests in scientific research, particularly the Commission of the European Communities and the Academia Europaea.

ESF pays for its activities through the contributions of its member organizations. Contributions are made both to the ESF basic budgets, to which member organizations contribute on a scale that is calculated on the basis of national incomes, and to specific programs and projects in which certain member organizations may have particular interest.

The modes of ESF work vary with the expressed needs and may change in character over the years. Broadly, ESF scientific programs almost always contain teams of scientists who carry out research. ESF scientific networks discuss, plan, innovate, analyze, or coordinate research, but seldom carry out large amounts of substantive research. Programs are often long-term and are funded (except in the developmental phase) by participating member organizations. Networks are usually of shorter term (three years) and funded from the network account within the ESF basic budget. In early 1993, there were 23 ESF scientific networks and 31 scientific programs in operation, with others being prepared. ESF also organizes, jointly with the Commission of the European Communities, a program of European research conferences.

New programs have increasingly been associated with the leading edge of science, such as the program (launched this year) in kinetic processes in minerals and ceramics. This aims to build a European bridge between geoscientists mainly concerned with minerals and minerals scientists working on inorganic substances, and to study the physical chemistry and the kinetics of processes.

Other programs under active discussion are

concerned with the increasingly important technical sciences, such as that on process integration in biochemical engineering (linking basic bioscience with possible industrial applications) and the mathematical treatment of free boundary problems (which is at the interface between engineering and the fundamental natural and mathematical sciences—such changing boundaries are found in material processing, biology, combustion theory, electrochemistry, and fluid flow).

The success of ESF in bringing together the natural and social sciences has been impressive, particularly in those areas, such as the study of environmental matters, where scientific research is vital in providing adequate and accurate evidence to underpin economic and political decisions affecting the world's population.

In line with the many challenging changes that have taken place in Europe over the last few years—and continue to take place—the nature of ESF's support for basic sciences has changed and perhaps has become more responsive to the demands of scientists in the Europe that is now taking shape.

With the transformation of Eastern Europe, the disintegration of some formerly monolithic states, and the decentralization of power to local (and sometimes disputatious) communities, there is a great need for institutions that bring together people of like interests and knowledge to work for a shared purpose and a common good.

## INTERNATIONAL COUNCIL OF SCIENTIFIC UNIONS

The International Council of Scientific Unions (ICSU) is a nongovernmental organization created in 1931 and a direct successor to the International Research Council established in 1919. The objectives of ICSU are straightforward: to encourage and promote international scientific and technological activity for the benefit and well-being of humanity. As a network of networks, ICSU achieves these objectives through a variety of ways, including the coordination of activities of its 20 scientific union members and its 86 national scientific members. ICSU also stimulates, designs, or participates in the implementation of international interdisciplinary programs and acts as a consultative body on international scientific issues. In pursuing its objectives, ICSU observes and actively upholds the principle of the universality of science.

In addition to the creation of interdisciplinary bodies on such topics as oceans, space science, global environment, biotechnology, natural disasters, and scientific data, one of the driving forces for joint action by the ICSU membership is the common concerns in which all members are stakeholders. An underlying theme in all of ICSU's work is the need to have strong scientific communities in all countries so that the maximum number can participate in national and international scientific actions. As the activities of ICSU become more global in nature, there is an ever growing need for well-trained scientists to contribute to these efforts. The need for increased capacity building in science was an important message from ICSU's ASCEND 21 meeting, which set down the agenda for science and development into the twenty-first century. This agenda was translated in Chapter 35, "Science for Sustainable Development," in Agenda 21 of the United Nations Conference on Environment and Development (Rio, 1992). One of its key recommendations was that by the year 2000 there should be a substantial increase in the number of scientists in developing countries that lack researchers and the exodus of scientists from developing regions should be reversed.

In 1990, the ICSU General Assembly passed a resolution that stated that while "we live in an era of unprecedented progress in science, the attraction of science to the younger generation seems to be lessening in some countries" and that available statistics "point to the danger of insufficient human resources in science and technology as the twenty-first century opens." The resolution asked ICSU, together with other concerned bodies, to examine the magnitude of this problem. Through conferences initiated by its members (in this case, the U.S. National Research Council), through a lectureship program sponsored by UNESCO, and through its Committee on Science and Technology in Developing Countries and Committee on Capacity Building in Science, ICSU hopes to contribute to the solution of these problems.

## NORTH ATLANTIC TREATY ORGANIZATION

In addition to its well-known political and military dimensions, the North Atlantic Treaty Organization (NATO) has a third dimension that seeks to encourage interaction between people, to consider some of the challenges facing our modern society, and to foster the development of science and technology. The programs of the NATO Science Committee are a major component of this third dimension.

The objective of the science program is the enhancement of science and technology through a variety of activities aimed at promoting international scientific cooperation. Most fields of science are eligible for support under general exchange programs of advanced study institutes, collaborative research grants, and science fellowships, while a number of special programs give support in a few specific areas of science considered to be in need of short-term concentrated effort.

Recently, links with the countries of Central and Eastern Europe, NATO's Cooperation Partners, have become an important aspect of the science program, in addition to the transatlantic link that has been and remains a major feature of these cooperative programs. The science committee is reorienting some of its activities to address scientific and technological problems being encountered by the Cooperation Partners, members of the North Atlantic Cooperation Council. Following consultation with representatives of the Cooperation Partner countries, a number of priority areas have been selected for support:

- Disarmament Technologies: scientific problems related to disposal of nuclear and chemical weapons, defense industry conversion, and safety of nuclear technologies.

- Environment: scientific problems related to reclamation of contaminated military sites, regional environmental problems, and natural and man-made disasters.

- High Technology: scientific problems related to information science, materials science, biotechnology and energy conservation, and supply (non-nuclear).

- Human Resources: problems related to science policy, management, intellectual property rights, and career mobility (e.g., the redeployment of defense industry scientists).

The NATO science program is managed by the staff of the NATO Scientific Affairs Division, under the overall policy guidance of the NATO Science Committee, with the assistance of a number of panels of scientists drawn from all countries of the alliance.

## ORGANIZATION FOR ECONOMIC COOPERATION AND DEVELOPMENT

The Organization for Economic Cooperation and Development (OECD), an intergovernmental organization that today comprises 24 democratic nations with advanced market economies, was founded in 1960 with the basic aim of promoting policies to:

- achieve the highest sustainable economic growth and employment;

- contribute to economic and social welfare throughout the OECD area;

- stimulate and harmonize its members' efforts in favor of developing countries; and

- contribute to the expansion of world trade on a multilateral nondiscriminatory basis.

Member countries are Australia, Austria, Belgium, Canada, Denmark, Finland, France, Germany, Greece, Iceland, Ireland, Italy, Japan, Luxembourg, the Netherlands, New Zealand, Norway, Portugal, Spain, Sweden, Switzerland, Turkey, the United Kingdom, and the United States. The Commission of the European Communities takes part in the work of the OECD.

Work on human resources in science and technology is undertaken by two subsidiary bodies of the Committee for Scientific and Technological Policy: the group on the science system dealing with policy aspects and the group of national experts of science and technology indicators working with the staff of the Economic Analysis and Statistics Division of the Directorate for Science, Technology, and Industry. It draws on the more general work of the Directorate for Education, Employment Labor, and Social Affairs.

Currently, there are no strictly comparable international statistics on the human resources used in industrial, technological, and scientific activities. This explains the importance of the work undertaken jointly with the Commission of the European Communities for the development of a new manual and new indicators based on the use of national and international education and employment statistics that already exist but do not yet allow reliable international comparisons. These new indicators of stocks and flows are to be used by authorities, universities, and industry to define supply

*PROFILES OF PARTICIPATING ORGANIZATIONS*

and demand concerning high-level staff more accurately, taking into account various factors such as population, aging, mobility (including brain drain), enrollment by women in higher education, and the numbers of new university graduates.

## THIRD WORLD ACADEMY OF SCIENCES

The Third World Academy of Sciences (TWAS) was founded in 1983 and officially launched and inaugurated by the Secretary General of the United Nations, Mr. Perez de Cuellar, in 1985. It has succeeded in uniting the most distinguished scientists from the Third World. Currently, it has 325 members from 54 Third World countries, including the 9 living science Nobel Laureates of Third World origin. The founding President of the Academy is the Nobel Laureate Professor Abdus Salam of Pakistan.

The Academy is a nongovernmental, nonpolitical, and nonprofit organization whose main objectives are to support scientific excellence and research in the Third World. It does this by awarding annual prizes to eminent scientists from the Third World who have made significant contributions in science, by providing fellowships to facilitate contacts between research workers in developing countries, and by encouraging scientific research on major Third World problems. The Academy was granted official nongovernmental consultative status with the United Nations Economic and Social Council in 1985. It is presently located on the premises of the International Center for Theoretical Physics at Miramare, Trieste, Italy, a center sponsored by the International Atomic Energy Agency and the United Nations Educational, Scientific and Cultural Organization.

In 1988, TWAS assisted in the establishment of the Third World Network of Scientific Organizations (TWNSO). TWNSO is a nongovernmental, nonprofit, and autonomous scientific organization whose objective is to enhance the economic development of the Third World and the cooperation among nations of the south in areas of science and technology critical to their sustainable development. The membership of TWNSO currently stands at 123, including 26 ministries, 42 research councils, and 34 academies in 69 Third World countries.

TWAS and TWNSO are currently promoting the establishment of a network of international centers in the south dedicated to human resource development. The first phase in the development of the network will aim at upgrading a number of existing competent national centers to international levels, to enable them to offer training and research opportunities to other scientists. A comprehensive feasibility study of the project is currently being undertaken in collaboration with the United Nations Industrial Development Organization (UNIDO) and the ICSU.

In addition, TWNSO, in collaboration with the south center and UNIDO, is compiling an inventory of world-class research and training institutions in various countries in the south with the purpose of facilitating the launching of a comprehensive fellowship program for scientists from the south who wish to pursue research and training in those centers. The first issue of the inventory will be published in September/October 1993 and will be circulated widely in the south and north.

## U.S. NATIONAL SCIENCE FOUNDATION

The National Science Foundation (NSF) is an independent federal agency created in 1950. Its aim is to promote and advance scientific progress in the United States. NSF has a unique role in the federal government: it is responsible for the overall health of science and engineering across all disciplines in contrast to other agencies that support research and development focused on their particular mission (e.g., health or defense). NSF is directed by a presidentially-appointed Director and Board of 24 scientists and engineers, including top university and industry officials. NSF is structured much like a university, with grant-making divisions for the various disciplines and fields of science and engineering.

NSF is also committed to expanding the United State's supply of scientists, engineers, and science educators, and has a variety of programs to encourage and enhance science and math education at all levels from precollege to graduate education and teacher training. The education and retention of women and minorities in science and engineering is an area of particular concern. NSF is also the federal agency responsible for collecting, analyzing, and disseminating quantitative data on the science and engineering enterprise and has a highly developed system of surveys related to the science and engineering pipeline and workforce. NSF publishes a number of congressionally mandated reports that cover human resource development topics, including *Science and Engineering Indicators, Women and Minorities in Science and Engineering,* and *Indicators of Science and Mathematics Education.*

# Appendixes

# Appendix A

## The Conference Program

TRENDS IN SCIENCE AND TECHNOLOGY CAREERS:
An International Conference
Palace Hotel
Brussels, Belgium

March 28-30, 1993

**Evening Before, March 28, 1993**

7:30 p.m.   Reception and Dinner

Welcoming Remarks: Paola Fasella
Director General, DG XII
Commission of the European Communities

Conference Overview: Walter A. Rosenblith, Chair

**March 29, 1993**

8:45 a.m.   Opening Remarks: Hywel Jones
Task Force on Human Resources
Commission of the European Communities

9:30 a.m.   **PANEL 1: MONITORING CAREER TRENDS IN SCIENCE AND TECHNOLOGY (S&T)**

Session Co-Chair:
**Richard Pearson**
Institute of Manpower Studies, University of Sussex (United Kingdom)

Session Co-Chair:
**Daniel Kleppner**
Massachusetts Institute of Technology (USA)

9:30 a.m.   Speaker 1:
**Gunnar Westholm**
Organisation for Economic Co-operation and Development (France)

10:00 a.m.  Speaker 2:
**Hajime Nagahama**
National Institute of Science and Technology Policy, Second Policy-Oriented Research Group (Japan)

## March 29, 1993

| | |
|---|---|
| 10:30 a.m. | Discussant:<br>**Glynis Breakwell**<br>University of Surrey (United Kingdom) |
| 11:00 a.m. | **General Discussion and Session Summary**<br>Discussion Leader: Session Co-Chair<br>**D. Kleppner** |
| 12:30 p.m. | Lunch |
| 2:00 p.m. | **PANEL 2: ANALYZING TRENDS IN S&T CAREERS: THE LONGITUDINAL APPROACH**<br><br>Session Co-Chair:<br>**Paul Baltes**<br>Max Planck Institute for Human Development and Education (Germany)<br><br>Session Co-Chair:<br>**Stephen Berry**<br>University of Chicago (USA) |
| 2:00 p.m. | Speaker 1:<br>**Yu Xie**<br>The University of Michigan (USA) |
| 2:30 p.m. | Speaker 2:<br>**Emilio Muñoz**<br>Institute of Advanced Social Studies (Spain) |
| 3:00 p.m. | Discussant:<br>**Jon Miller**<br>The Chicago Academy of Sciences (USA) |
| 3:30 p.m. | **General Discussion and Session Summary**<br>Discussion Leader: Session Co-Chair<br>**S. Berry** |
| 4:00 p.m. | Break |
| 4:30 p.m. | **PANEL 3: ANALYZING TRENDS IN S&T CAREERS: FACTORS DETERMINING CHOICE**<br><br>Session Co-Chair:<br>**Wendy K. Hansen**<br>Department of Industry, Science and Technology (Canada)<br><br>Session Co-Chair:<br>**John Moore**<br>George Mason University (USA) |
| 4:30 p.m. | Speaker 1:<br>**Tom Whiston**<br>Science Policy Research Unit, University of Sussex (United Kingdom) |
| 5:00 p.m. | Speaker 2:<br>**Torsten Husèn**<br>Institute of International Education, Stockholm University (Sweden) |
| 5:30 p.m. | Discussant:<br>**Alfred McLaren**<br>Science Service, Inc. (USA) |
| 6:00 p.m. | **General Discussion and Session Summary**<br>Discussion Leader: Session Co-Chair<br>**J. Moore** |
| 6:30 p.m. | Adjourn for Dinner at 7:30 p.m.<br>**Dinner Speaker:**<br>**Marianne Frankenhaeuser** |

## March 30, 1993

| | |
|---|---|
| 8:30 a.m. | Conference Reconvenes:<br>Walter Rosenblith, Chair |
| 8:45 a.m. | Opening Remarks: Human Capital and Mobility Program<br>Commission of the European Communities |

*APPENDIX A*

| | | | |
|---|---|---|---|
| 9:00 a.m. | **PANEL 4: UTILIZING POINTS OF INTERVENTION TO ENHANCE AND SUSTAIN INTEREST IN S&T CAREERS** | 2:00 p.m. | **PANEL 5: INFLUENCING S&T CAREER TRENDS: THE ROLE OF INTERNATIONAL ORGANIZATIONS** |

Session Co-Chair:
**Pim Fenger**
Ministry of Education and Science
(Netherlands)

Session Co-Chair:
**Ernest Jaworski**
Monsanto, Inc. (USA)

9:00 a.m.  Speaker 1:
**Kazuo Ishizaka**
National Institute for Educational Research (Japan)

9:30 a.m.  Speaker 2:
**Dervilla Donnelly**
University College Dublin (Ireland)

10:00 a.m.  Discussant:
**Pamela Ebert Flattau**
National Research Council (USA)

10:30 a.m.  **General Discussion and Session Summary**
Discussion Leader: Session Co-Chair
**E. Jaworski**

12:30 p.m.  Lunch

Session Co-Chair:
**Rodney Nichols**
New York Academy of Sciences
(USA)

- Academia Europaea (Sir Arnold Burgen)
- Commission of the European Communities
- European Science Foundation (D. Donnelly)
- International Council of Scientific Unions (J. Marton-Lefevre)
- North Atlantic Treaty Organization (J-M Cadiou)
- Petrofina (J. Rebelo)
- Organization for Economic Cooperation and Development (D. Malkin)
- Third World Academy of Sciences (C. Ponnamperuma)
- UN Educational, Scientific and Cultural Organization (C. Power)
- U.S. National Science Foundation (C. Marrett)

4:30 p.m.  **General Discussion and Session Summary**
Discussion Leader: Session Co-Chair
**R. Nichols**

6:00 p.m.  **Closing Remarks:**
**Walter Rosenblith**
Adjournment

# Appendix B

## Participant Roster

Sherburne ABBOTT
Committee on International
  Organizations and Programs
National Academy of Sciences
Washington, DC  USA

Paul BALTES
Max Planck Institute for Human Development
 and Education
Berlin, GERMANY

Catherine BEC
ARNT
Paris, FRANCE

Daniel De BERGHERS
Commission of the European Communities
Plateau de Kirschberg, Luxembourg

Stephen BERRY
The James Franck Institute
University of Chicago
Chicago, Illinois  USA
*(as of March 1993)*

Glynis BREAKWELL
Department of Psychology
University of Surrey
Guldford, UNITED KINGDOM

Guiliana BURESTI
Commission of the European Communities
Brussels, BELGIUM

Sir Arnold BURGEN
Academia Europaea
London, UNITED KINGDOM

Jean-Marie CADIOU
North Atlantic Treaty Organization
Brussels, BELGIUM

Dervilla DONNELLY
University College Dublin
Dublin, IRELAND

Annalise EGGIMANN
Swiss National Science Foundation
Bern, SWITZERLAND

Paolo FASELLA
Commission of the European
  Communities
Brussels, BELGIUM

Gilbert FAYL
Science and Technology
Delegation of the Commission of
 the European Communities
Washington, DC  USA

Pim FENGER
Ministry of Education and Science
Zoetermeer, NETHERLANDS

*APPENDIX B*

Pamela Ebert FLATTAU
Office of Scientific and Engineering
  Personnel
National Academy of Sciences
Washington, DC  USA

Marianne FRANKENHAEUSER
Karolinska Institutet
University of Stockholm
Stockholm, SWEDEN

Francois GOVAERTS
Cooperation with Industrialized
  Countries Outside of Europe
Commission of the European
  Communities
Brussels, BELGIUM

Wendy HANSEN
Scientific, Technical and Engineering
  Personnel
Industry, Science, and Technology
Ottawa, CANADA

Mohamed Hag Ali HASSAN
Third World Academy of Sciences
International Center for Theoretical
  Physics
Trieste, ITALY

Torsten HUSÉN
Institute of International Education
Stockholm University
Stockholm, SWEDEN

Kazuo ISHIZAKA
National Institute for Educational
  Research
Tokyo, JAPAN

Ernest JAWORSKI
Distinguished Science Fellow
11 Cleerbrook Lane
St. Louis, Missouri  USA

Hywel JONES
Task Force on Human Resources
Commission of the European
  Communities
Brussels, BELGIUM

Daniel KLEPPNER
Department of Physics
Massachusetts Institute of Technology
Cambridge, Massachusetts  USA

Patrice LEGRO
Committee on International
  Organizations and Programs
National Academy of Sciences
Washington, DC  USA

Hugh LOGUE
Commission of the European Communities
Brussels, BELGIUM

Alfred McLAREN
Science Service, Inc.
Washington, DC  USA

Daniel MALKIN
Head of the Economic Analysis
  and Statistics Division
Organization for Economic Cooperation
  and Development
Paris, FRANCE

Cora MARRETT
Directorate for Social, Behavioral, and
  Economic Science
National Science Foundation
Washington, DC  USA

Julia MARTON-LEFÈVRE
International Council of Scientific
  Unions
Paris, FRANCE

Lene MEJER
Statistical Office of the European Communities
Commission of the European Communities
Brussels, BELGIUM

Jon MILLER
The Chicago Academy of Sciences
International Center for the Advancement of Scientific
  Literacy
Chicago, Illinois  USA

John MOORE
International Institute
George Mason University
Arlington, Virginia USA

Emilio MUÑOZ
Institute of Advanced Social Studies
Madrid, SPAIN

Jean Dreux de NETTANCOURT
Human Capital and Mobility Program
Commission of the European
  Communities
Brussels, BELGIUM

Rodney NICHOLS
New York Academy of Sciences
New York, New York USA

Frank OWENS
Office of Human Resources and
  Education
National Aeronautics and Space
  Administration
Washington, DC USA

Richard PEARSON
Institute of Manpower Studies
University of Sussex
Brighton UNITED KINGDOM

Ian PERRY
Directorate-General DG XII
for Science, Research and Development
Commission of the European
  Communities
Brussels, BELGIUM

Vincent PIKET
COMETT
Commission of the European Communities
Brussels, BELGIUM

Paul RAMBAUT
Scientific and Environmental Affairs
  Division
North Atlantic Treaty Organization
Brussels, BELGIUM

Galo RAMIREZ
Universidad Autónoma de Madrid
Centro de Biologia Molecular
Cantoblanco,
Madrid, SPAIN

Jose REBELO
Human Resources Department
Petrofina
Brussels, BELGIUM

Walter ROSENBLITH
Massachusetts Institute of Technology
Cambridge, Massachusetts USA

Ryuji SHIMODA
1st Policy-oriented Research Group
National Institute of Science and
  Technology Policy
Tokyo, JAPAN

Lily TALAPESSY
Commission of the European Communities
Brussels, BELGIUM

Gunnar WESTHOLM
Organisation for Economic Co-operation
  and Development
Paris, FRANCE

Thomas WHISTON
Science Policy Research Unit
University of Sussex
Brighton UNITED KINGDOM

Charles WHITE
Commssion of the European
  Communities
Brussels, BELGIUM

Yu XIE
The University of Michigan
Population Studies Center
Ann Arbor, Michigan USA

Alison YOUNG
Organisation for Economic Co-operation
  and Development
Paris, FRANCE